Wastewater Engineering and Treatment 3

Wastewater Engineering and Treatment 3

Elimination of Nitrogen and Phosphorus, Clarification

Kader Gaid

iSTE

WILEY

First published 2026 in Great Britain and the United States by ISTE Ltd and John Wiley & Sons, Inc.

ISTE Ltd
27-37 St George's Road
London SW19 4EU
UK

www.iste.co.uk

John Wiley & Sons, Inc.
111 River Street
Hoboken, NJ 07030
USA

www.wiley.com

Library of Congress Control Number: 2026936291

British Library Cataloguing-in-Publication Data
A CIP record for this book is available from the British Library
ISBN 978-1-83669-057-3

The manufacturer's authorized representative according to the EU General Product Safety Regulation is Wiley-VCH GmbH, Boschstr. 12, 69469 Weinheim, Germany, e-mail: Product_Safety@wiley.com.

Contents

Chapter 1. Biological Removal of Nitrogen 1

1.1. Nitrogen sources 2
1.2. Biochemical mechanisms of nitrogen utilization 4
1.2.1. Ammonification 4
1.2.2. Nitrification 6
1.2.3. Denitrification 9
1.3. Factors influencing nitrification 16
1.3.1. Temperature 17
1.3.2. Growth rate 18
1.3.3. Oxygen 19
1.3.4. pH and alkalinity 19
1.3.5. Nutrients 20
1.3.6. Dissolved oxygen and nitrogen substrate 20
1.4. Nitrification in an activated sludge system 22
1.4.1. Influence of mass loading 22
1.4.2. Influence of TSS/MVS concentration in the tank 23
1.4.3. Influence of loading rate 23
1.4.4. Influence of sludge age 24
1.4.5. Nitrification rate 24
1.4.6. O_2 requirement 25
1.4.7. Dynamic model of nitrification 27
1.5. Denitrification in activated sludge and denitrification rate 30
1.5.1. Specific anoxic zone for denitrification 34
1.5.2. Estimation of excess sludge production 35
1.5.3. Carbon addition 36
1.6. Nitrification and denitrification processes in activated sludge 37
1.6.1. Pre-denitrification 37
1.6.2. Post-denitrification 37

1.7. Autotrophic denitrification of activated sludge 40
1.7.1. Autotrophic denitrification mechanism with sulfur 41
1.7.2. Autotrophic denitrification in free culture (activated sludge) 43
1.7.3. Autotrophic denitrification in a biofilter 46
1.8. Nitrogen treatment by anammox bacteria 52
1.8.1. Anammox characteristics. 54
1.8.2. Microbiological characteristics 54
1.8.3. Biochemistry of anaerobic oxidation of ammonium 55
1.8.4. Factors influencing the growth of anammox bacteria 57
1.8.5. Salinity. 57
1.8.6. Presence of organic matter and/or nutrients 57
1.8.7. *Candidatus Brocadia* and *Kuenenia*. 58
1.8.8. Oxygen. 60
1.8.9. Temperature 60
1.8.10. Substrate concentrations 60
1.8.11. Nitrite. 60
1.8.12. Ammonium ion (NH_4^+) 61
1.8.13. Inhibition mechanisms and pH 62
1.8.14. Implementation using anammox bacteria 62
1.8.15. Nitrogen removal processes using anammox 67
1.8.16. Applications 76
1.9. MABR: a new opportunity 79
1.10. Conclusion 80
1.11. References 81

Chapter 2. Phosphorus Removal 93

2.1. The phosphorus cycle 93
2.2. Eutrophication 95
2.3. Oxidized mineral forms of phosphorus in wastewater 96
2.4. Phosphorus removal processes 100
2.4.1. Overview 100
2.4.2. Biological phosphorus removal 101
2.5. Treatment processes incorporating biological dephosphatation 116
2.5.1. Activated sludge 117
2.5.2. Biofilters: introduction of reagents into the primary settling tank. 125
2.5.3. Other biological phosphorus removal technologies 125
2.6. Selectors 130
2.6.1. Aerobic selector. 131
2.6.2. Anoxic selector 131
2.6.3. Anaerobic selector 132

2.7. Physicochemical treatment 133
2.7.1. Fundamental principles of phosphorus precipitation. 135
2.8. Implementation of chemical reagents 152
2.8.1. Injection points 152
2.8.2. Injection control. 152
2.9. Other technical solutions for phosphorus removal 156
2.9.1. Active media for phosphorus removal 156
2.9.2. Artificial wetlands 159
2.9.3. Ion exchange 159
2.10. Overall phosphorus removal performance 160
2.11. References 161

Chapter 3. Clarification 171

3.1. Introduction 171
3.1.1. Principle 171
3.1.2. Types of clarifiers. 172
3.2. Settling rate of particles in a liquid. 173
3.2.1. Diluted suspensions. 173
3.2.2. Flocculated particles 176
3.2.3. Concentrated suspensions 177
3.2.4. Sedimentation curves. 178
3.3. Settling theory 180
3.3.1. Fundamental law: Kynch theory. 180
3.4. Mathematical interpretation of the different types of settling 187
3.4.1. Free or coalescent settling (zone B, Figure 3.2) 187
3.4.2. Hindered settling (zone C, Figure 3.2) 193
3.4.3. Compaction zone 197
3.5. Sizing of clarifiers 197
3.5.1. Different sizing approaches 201
3.5.2. Coalescing settling (zone B, Figure 3.13). 203
3.5.3. Hindered settling zone and sludge blanket clarifiers (zone C, Figure 3.13) 207
3.5.4. Sludge compaction zone (zone D, Figure 3.13) 208
3.5.5. Relationship between settling rate and height of equipment 211
3.6. Characteristics of classic clarifiers 215
3.6.1. Horizontal clarifiers. 216
3.6.2. Cylindrical clarifiers 221
3.6.3. Other technologies 225
3.6.4. General design of conventional clarifiers 225

3.7. Inclined pipe clarifiers 232
3.7.1. Theory and principle 233
3.7.2. Sizing principles for lamellar clarifiers 236
3.8. Clarifier malfunctions 243
3.9. Technologies 245
3.9.1. Clariflo® clarifier (Veolia) 245
3.9.2. Rim-Flo® clarifier 247
3.9.3. TowBro® scraper 249
3.9.4. Lamellar settling tank: Multiflo® (Veolia) 250
3.10. References 255

Index 259

1

Biological Removal of Nitrogen

Various measures have been taken over the last two decades by administrative authorities of many countries in the attempt to combat the increasing pollution of surface waters, a significant consequence of which can be eutrophication.

> Eutrophication is enrichment of nutrients in waters that stimulates an array of symptomatic changes, which can include increased phytoplankton and rooted aquatic plant production, fisheries and water quality deterioration, and other undesirable changes that interfere with water uses. (Pannard et al. 2017)

Eutrophication is an extremely complex phenomenon but can be described schematically: wastewater containing large quantities of nutrients such as nitrogen and phosphorus, when discharged into receiving environments, promotes the rapid and continuous growth of algae and aquatic plants.

Algae have a very short lifespan and decompose rapidly. This organic decomposition results in a high oxygen demand in the environment. The oxygen deficit creates anaerobic conditions, the initial consequences of which are the formation of hydrogen sulfide, ammonia, mineral matter, etc., which gradually clog the receiving environment where wastewater is discharged.

Of the nutrients responsible for these phenomena, nitrogen and phosphorus are the most detrimental. The two main methods of controlling eutrophication are the removal of nitrogen and the removal of phosphorus.

1.1. Nitrogen sources

Nitrogen is present in the air, soil and water in various forms, the chemical and biological combinations of which lead to very different reaction products, allowing us to deduce the cycle shown in Figure 1.1.

Nitrogen is found in mineral and organic forms. Mineral nitrogen is primarily in the form of ammonium ions (NH_4^+), nitrite ions (NO_2^-) and nitrate ions (NO_3^-). Organic nitrogen is present in the form of proteins which, through hydrolysis, form amino acids which in turn, through condensation, form peptides and polypeptides. It is also found in a wide variety of other organic compounds (urea, uric acid, creatinine, etc.).

Nitrogen removal presents serious ecological challenges for receiving water bodies, including eutrophication and toxicity to aquatic life due to its dissolved form as various compounds (Wiesmann 1994; Wang et al. 2017).

When organic matter is broken down by microorganisms, organic nitrogen is transformed into NH_4^+. The main wastewater concentrations of NH_4^+ and organic nitrogen are approximately 40 and 20 mg $N.L^{-1}$, respectively (Wiesmann 1994).

Nutrients contained in all kinds of organic matter can be recycled in various forms. Livestock effluents, crop residues and other organic by-products of human activities are a large resource for fertilization. Only prokaryotic organisms can reduce nitrogen into an assimilable recombinant form. Rhizobia belong to this group. The most efficient nitrogen-fixing systems are symbioses that couple nitrogen fixation with photosynthesis.

Ammonia synthesis uses natural gas and water to produce hydrogen gas (H_2) and combines it with nitrogen (N_2) present in the atmosphere to form ammonia (NH_3), a key ingredient in major nitrogen fertilizers.

Nitrogen fixation is the conversion of atmospheric nitrogen into nitrogen usable by plants and animals. This is carried out by certain bacteria that live in soils or water and can assimilate diatomic nitrogen N_2 and use it to synthesize their proteins.

Nitrogen reduction leads to the formation of diimine, and its oxidation leads to nitrous oxide (N_2O).

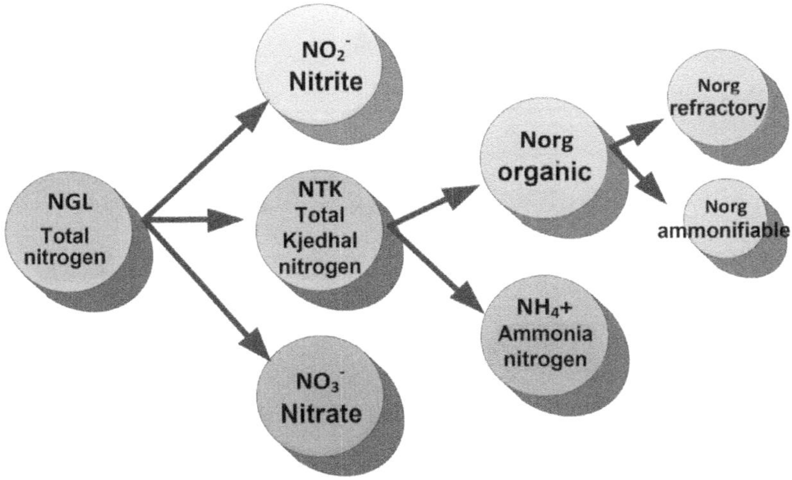

Figure 1.1. *Forms of nitrogen in wastewater*

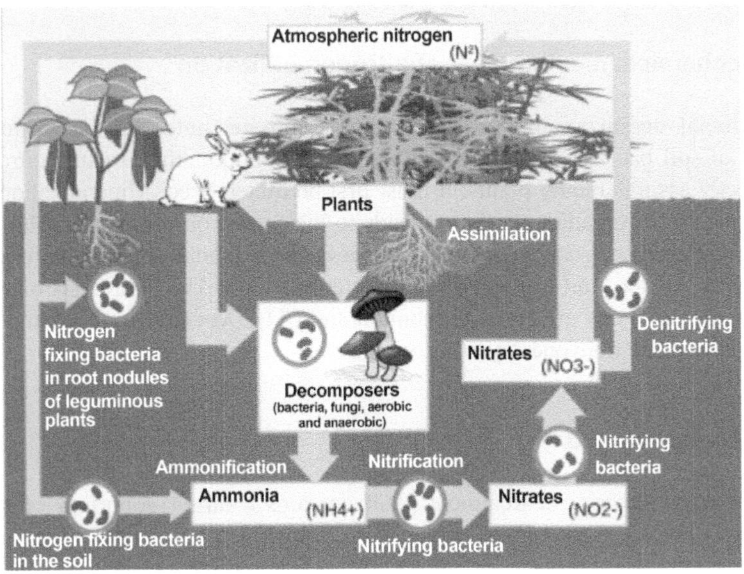

Figure 1.2. *Nitrogen cycle*

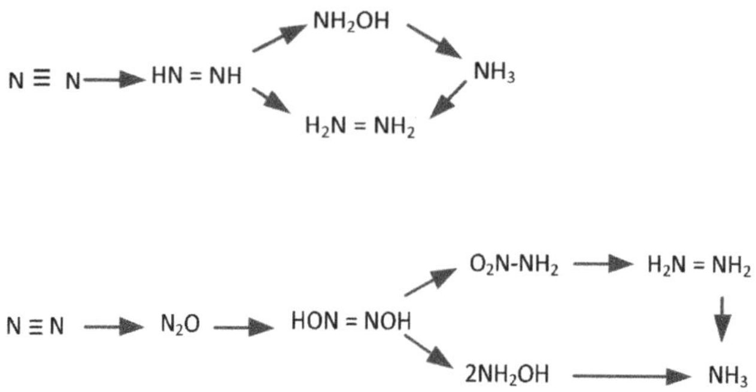

a)

b)

Figure 1.3. *a) Diimine. b) N_2O as intermediary*

Nitrogen fertilizer can contain nitrogen in the form of urea, ammonia, nitrate or a mixture of these forms. It is available in liquid form (nitrogen solution) or granular form (urea, ammonium nitrate).

1.2. Biochemical mechanisms of nitrogen utilization

Microbial organization transforms mineral nitrogen into organic matter. The activity of soil bacteria is primarily stimulated by ammonium. Organic nitrogen is not directly assimilable by plants; it must first be mineralized. Mineralization is the transformation of organic matter that leads to the formation of mineral salts, where fertilizing elements become soluble and accessible to plants. The mineralization of soil organic matter (and effluents) produces ammonium. The hydrolysis of urea by soil enzymes converts urea into ammonium and CO_2. Depending on the temperature, hydrolysis can take from one day to one week.

1.2.1. *Ammonification*

Wastewater entering a treatment plant contains a large proportion of organic nitrogen (albumin, carbamide). Furthermore, nitrogen discharged from residential settings is initially slightly more concentrated in ammonifiable organic nitrogen.

Organic nitrogen is said to be ammonifiable when it can be transformed by enzymatic hydrolysis into ammoniacal nitrogen. In other words, the ammonifiable/ammoniacal ratio can depend on factors such as temperature and incubation time (depending on the length of the network) because there is a certain point that organic nitrogen is transformed into N-NH_4^+.

During treatment in a wastewater treatment plant, ammonification continues until the majority of the nitrogen is transformed into N-NH_4^+.

Soluble refractory nitrogen is the non-biodegradable component of nitrogen. Furthermore, this component is already detected upon entering the treatment plant. However, the quantity of this nitrogen in standard urban wastewater ranges from 1.5 to 2.5 $mg.L^{-1}$. Hard or refractory particulate nitrogen has similar concentrations but will be trapped in sludge. In some cases, the concentration of soluble hard nitrogen may be higher. This is generally due to sludge return or a system with a long retention time.

Ammonification involves nitrogenous substances of very diverse structures: proteins, amino sugars, nucleic acids, purine bases, amides, phosphatides, amines, uric acid and urea. The ammonia released from a protein substance corresponds to the following formula:

$N(NH_3)$ = N protein – (NH_3 metabolized by heterotrophs + NH_3 oxidized by autotrophs)

Ammonium is absorbed by roots more slowly than nitrate. This is particularly true for cyanobacteria and certain bacteria living in symbiosis with plants (including legumes). The typical chemical reaction is:

$$2\,N_2\,(g) + 3\,CH_2O + 3\,H_2O \rightarrow 4\,NH_4^+ + 3\,CO_2$$

nitrogen (gas) — *organic matter* — *water* — *ammonia* — *carbon dioxide*

In soils with a high pH, ammonium is transformed into ammonia:

$$NH_4^+ + OH^- \rightarrow NH_3\,(g) \rightarrow H_2O$$

ammonium — *hydroxide* — *ammonia (gas)* — *water*

The reaction requires an energy input from photosynthesis (cyanobacteria and legume symbionts). This fixation tends to produce ammonia compounds such as ammonium NH_4^+ and its conjugate acid, ammonia NH_3. This is a reduction reaction that takes place via organic substances denoted $\{CH_2O\}$.

Positively charged ammonium (NH_4^+) is adsorbed into the soil (soil cation exchange capacity (CEC)), which exchanges it with the roots. Most of the ammonium is converted to nitrate before being absorbed by plants.

Nitrate is rapidly absorbed due to its high mobility in soil water solution. Most annual crops preferentially use nitrate over ammonium.

Ammonification is the transformation of organic nitrogen into ammoniacal nitrogen. This reaction largely takes place in the sewer system upstream of the wastewater treatment plant.

The residual organic nitrogen is removed from the plant in various ways: part of it is trapped by sedimentation in the primary treatment zone, another part by filtration in the biocarbon filter, and the remainder is transformed into ammoniacal nitrogen as it flows through the treatment plant, mainly at the primary settling point. Generally speaking, almost all organic nitrogen is removed downstream of the plant.

Assimilation occurs during the biological elimination of carbon pollution by bacterial cells, mobilizing nitrogen.

When microorganisms cease their biological activity, the lysis of bacterial membranes provides a source of organic matter hydrolyzed into carbon and nitrogen. Heterotrophic biomass, by far the most significant contributor, plays a crucial role in producing these mineral elements, the quantities of which are proportional to its concentration.

The mineralization of living biomass leads to 0.09 g $N.g^{-1}$ COD eliminated; that is, 0.12 g $N.g^{-1}$ MVS eliminated:

$$C_5H_7NO_2 + 5\ O_2 \rightarrow 5\ CO_2 + NH_3 + 2\ H_2O$$

To remove nitrogen from wastewater, it is necessary to oxidize ammonium to nitrate, which must then be reduced to nitrogen gas. The nitrogen gas can then be permanently removed from the wastewater.

The oxidation of ammonium to nitrate is called nitrification. Nitrification by soil bacteria transforms ammonium into nitrate within a period ranging from a few days to a few weeks. Losses in the form of nitrous oxide or nitrogen oxide may occur during this process.

1.2.2. *Nitrification*

Nitrification transforms the products of fixation (NH_4^+, NH_3) into NO_x (NO_2^- and NO_3^-), nitrites and nitrates.

This is an oxidation reaction catalyzed by bacteria in soils and water. The biology of nitrifiers is dominated by their autotrophic nature, and they derive all their energy from oxidation reactions. This energy allows them to reduce carbon dioxide and use the carbon it contains. They are completely independent of any organic matter.

Nitrifying bacteria are classified into two categories:

– *Nitrosomonas*, which oxidize ammonia into nitrites;

– *Nitrobacters*, which oxidize nitrites into nitrates.

Each genus is specific in its reactions, and together they carry out nitrification. There are some heterotrophs such as *Aspergillus flavus* which oxidize NH_4^+ not only into nitrites but also into nitrates. However, their activity is less significant than that of nitrifying bacteria.

Nitritation is an oxidation reaction that produces:

$$NH_4^+ \rightarrow NO_2^- \rightarrow NO_3^-$$

$$2\ NH_4^+ + 3O_2 \rightarrow 2\ NO_2^- + 2\ H_2O + 4\ H^+$$

The formation of bacterial cells is governed by the following reaction:

$$5\ CO_2 + NH_4^+ + 2H_2O \rightarrow C_5H_7NO_2 + 5\ O_2 + H^+$$

Combining these two equations allows us to write the general equation:

$$55\ NH_4^+ + 5C0_2 + 760_2 \rightarrow C_5H_7NO_2 + 54\ NO_2 + 52H_2O + 109H^+$$

The nitration reaction involves Nitrobacter bacteria, which oxidize nitrites to nitrates according to the following reaction mechanism:

$$NO_2^- \rightarrow NO_2 + e^-$$

$$NO_2^- + H_2O \rightarrow NO_3^- + 2H^+ + e^-$$

$$1/O_2 + 2\ H^+ + 2e^- \rightarrow H_2O$$

$$NO_2^- + 1/2O_2 \rightarrow NO_3^-$$

Combining the equations allows us to write the general equation:

$$400\ NO_2^- + 5\ CO_2 + NH_4^+ + 195\ O_2 + 2\ H_2O \rightarrow C_5H_7NO_2 + 400\ NO_3^- + H^+$$

The process is a two-step process. The reactions are as follows:

$$NH_4^+ + \frac{3}{2}O_2 \rightarrow NO_2^- + 2H^+ + H_2O$$
$$NO_2^- + \frac{1}{2}O_2 \rightarrow NO_3^-$$
$$\overline{NH_4^+ + 2O_2 \rightarrow NO_3^- + H_2O}$$

Under specific conditions, some bacteria are capable of carrying out both of the above reactions. In the general context of wastewater and the conditions under which the processes are implemented, nitrification results from the sequential action of two groups of nitrifying organisms. These groups of nitrifying bacteria are presented in Table 1.1. *Nitrosomonas* and *Nitrobacter* are normally designated as the main nitrifying bacteria in wastewater treatment

Oxidation of ammonia	Nitrite oxidation
Nitrosomonas	*Nitrobacter*
Nitrococcus	*Nitrospina*
Nitrospira	*Nitrococcus*
Nitrosolobus	*Nitrospira*
Nitrosovibrio	

Table 1.1. *Some bacteria responsible for nitrification*

Carbon dioxide is used as a carbon source to produce new cellular tissue. In other words, they are chemolithotrophs. Taking into account the production of cellular tissue, the overall reaction for the nitrification process becomes:

$$NH_4^+ + 1.86O_2 + 1.98HCO_3^- \rightarrow 0.020C_5H_7NO_2 + 0.98NO_3^- + 1.04H_2O$$

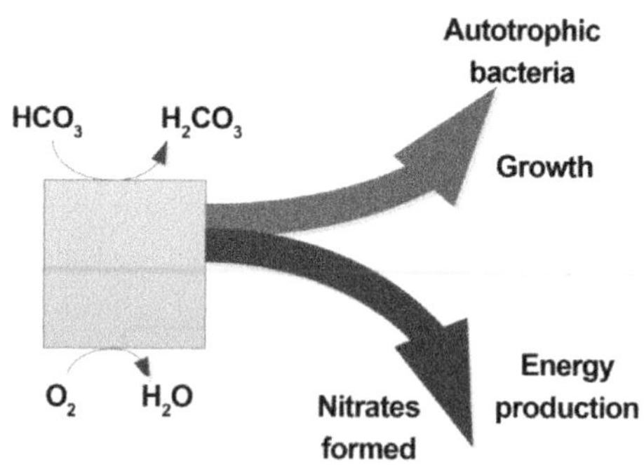

Figure 1.4. *Nitrogen utilization during assimilation (Spanjers 1996)*

This reaction shows that the nitrification process consumes both oxygen and alkalinity. The consumption of alkalinity means a decrease in pH. A significant decrease in pH will inhibit the process. An alkalinity of approximately 4 to 5 $meq.L^{-1}$ in wastewater is often necessary to support the nitrification process.

The nitrification of 20 $mg.L^{-1}$ d'N-NH_3 requires 84 $mg.L^{-1}$ dissolved oxygen and produces 3 $mg.L^{-1}$ with *Nitrosomonas*, 0.5 $mg.L^{-1}$ with Nitrobacter and 2 moles of H^+ per mole of oxidized ammoniacal nitrogen.

1.2.3. *Denitrification*

The process of reducing nitrites to nitrogen is called denitrification. Wastewater treatment plants can be designed with a variety of features combining nitrification and denitrification, depending on the desired level of removal. Denitrification occurs when microorganisms lack oxygen (stagnant water and compacted soil).

During this process, soil bacteria convert nitrates (and nitrites) into nitrogen gas (N_2) and, to a lesser extent, nitrous oxide (N_2O) and nitrogen oxides (NO_X), which are released into the atmosphere. Denitrification can be carried out by autotrophs and heterotrophs.

1.2.3.1. *Heterotrophic denitrification*

Denitrification returns nitrogen to the atmosphere in its molecular form N_2, with CO_2 as a by-product of nitrogen oxide N_2O, a greenhouse gas that contributes to the destruction of the ozone layer in the stratosphere. This is a reduction reaction of NO_3^- through bacteria that transform organic matter.

The reaction is as follows:

$$4NO_3^- + 5CH_2O + 4H^+ \rightarrow 2N_2\,(g) + 5CO_2\,(g) + 7H_2O$$

Two mechanisms of nitrate metabolism are possible: the dissimilatory pathway or the ammonification of nitrates:

$$NO_3^- \rightarrow NO_2^- \rightarrow NH_4^+ \text{ (by reductases)}$$

This reaction takes place in anaerobic habitats such as anoxic sediments and sludge, through strict or facultative anaerobic bacteria.

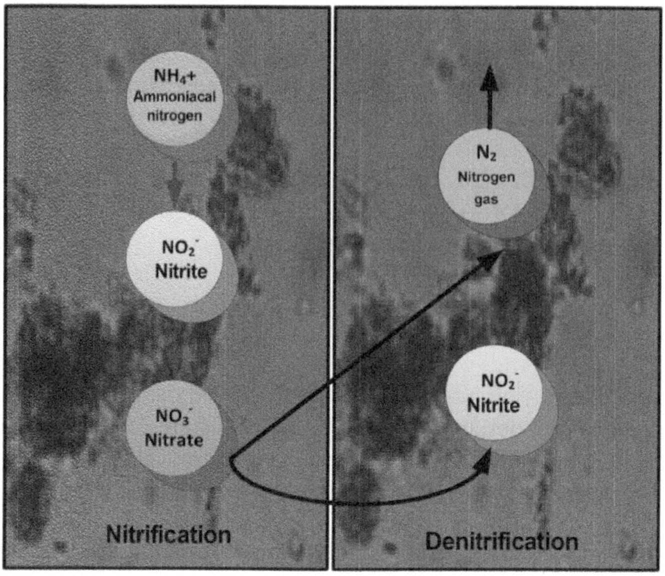

Figure 1.5. *Principle of biological elimination of mineral nitrogen*

Under certain conditions, microorganisms use nitrates as an oxidant instead of oxygen. The nitrates are then reduced to nitrogen and nitrogen oxides.

The goal of biological denitrification is therefore to completely remove nitrogen from wastewater. During this treatment process, nitrogen evaporates into the atmosphere in its molecular form, N_2. Denitrification is an anaerobic mechanism that allows a large number of heterotrophic bacteria to meet their energy needs using nitrates when dissolved oxygen is lacking. Without O_2, these bacteria must use the oxygen contained in nitrates for their respiratory needs.

The bacteria involved in the denitrification cycle are also involved in carbon alteration. In wastewater treatment plants, it is common to add an external carbon source (methanol, acetic acid, untreated sewage, industrial effluent) to compensate for carbon deficiencies.

Another advantage of denitrification is the recovery of alkalinity (a portion of that lost during the nitrification step). In fact, denitrification restores alkalinity equal to half the amount consumed by nitrification. That is, 1 kg of denitrified nitric nitrogen is equivalent to the addition of 1.95 kg of quicklime (CaO).

Furthermore, it is beneficial to implement a recirculation system in the water treatment process to reuse this alkalinity.

In conventional systems, biomass (activated sludge) is exposed to different environmental conditions to achieve COD and nitrogen removal.

Specifically, aerobic conditions are required for nitrification, while a certain amount of organic carbon compounds and the absence of oxygen are necessary to support denitrification.

The genera *Pseudomonas fluorescens* and *Bacillus* are the most common bacteria involved in heterotrophic denitrification. However, bacteria from other genera are possible: *Achromobacter*, *Aerobacter*, *Bacillus*, *Micrococcus*, *Paracoccus*.

Achromobacter	*Escherichia*	*Paracoccus*
Acinetobacter	*Flavobacterium*	*Propionibacterium*
Agrobacterium	*Gluconobacter*	*Pseudomonas*
Alcaligens	*Holobacterium*	*Rhizobium*
Bacillus	*Hyphomicrobium*	*Rhodopseudomonas*
Chromobacterium	*Kingella*	*Spirillum*
Cornybacterium	*Methanonas*	*Thiobacillus*
Denitrobacillus	*Moraxella*	*Xanthomonas*
Enterobacter	*Neisseria*	

Table 1.2. *Non-exhaustive list of heterotrophic bacteria in activated sludge (Ni et al. 2016)*

Numerous genera of bacteria, including heterotrophic denitrifying species, have been identified in activated sludge (see Table 1.2). Depending on the characteristics of the untreated water, treatment conditions and operating conditions, heterotrophic denitrifying bacteria represent approximately 50 to 80% of all flocculated and dispersed bacteria in activated sludge. Most of these denitrifying bacteria are facultative aerobic bacteria that prefer oxygen molecules to nitrogen oxide molecules as electron acceptors. Therefore, the absence of oxygen or the presence of an oxygen gradient within the activated sludge is necessary for efficient heterotrophic denitrification.

In this type of process, the source of organic carbon is fundamental: it is available in food and animal or plant waste. In the field of wastewater treatment, a carbon source is provided by organic carbon (BOD) already present in wastewater or through the addition of acetic acid, methanol or ethanol. These facultative anaerobic bacteria are capable of reducing nitrate and nitrite to nitrogen gas under anaerobic conditions. These microorganisms use either oxygen or oxidized forms of nitrogen as final electron acceptors.

With too little carbon, the reaction is incomplete and nitrites appear, while too much carbon promotes an overly reducing environment, which leads to the production of sulfides (rotten egg smell).

Industrial effluents laden with organic matter, such as those from refined sugar, fruit juice or even dairy products, can aid in denitrification.

The fundamental reaction in the presence of organic matter in wastewater is written as:

$$0.61.\ C_{18}H_{19}O_9N + 4.5\ NO_3^- + 0.39\ NH_4^+ + 4.15\ H_3O^+ \rightarrow C_5H_7NO_2 + 2.27\ N_2 + 5.98\ CO_2 + 5.15\ H_2O$$

This reaction of organic matter degradation, all other conditions being equal, is slower than that which occurs in the presence of oxygen. It is even slower the less readily biodegradable the available carbon.

The fundamental reactions of denitrification in the presence of methanol are:

– Required energy:

$$2CH_3OH + 6\ NO_3^- \rightarrow 2CO_2 + 4\ H_2O + 6\ NO_2^-$$

$$3CH_3OH + 6\ NO_2^- \rightarrow 3N_2 + 3CO_2 + 3H_2O + 6\ OH^-$$

i.e.:

$$5CH_3OH + 6\ NO_3^- \rightarrow 5\ CO_2 + 7\ H_2O + 3\ N_2 + 6\ OH^-$$

– Cell synthesis:

$$100\ CH_3OH + 97\ NO_3^- \rightarrow 5\ C_5H_7NO_2 + 75\ CO_2 + 134\ H_2O + 46\ N_2 + 97\ OH^-$$

– Endogenous respiration:

$$5C_5H_7NO_2 + 23NO_3^- \rightarrow 25CO_2 + 6H_2O + 14N_2 + 23OH^-$$

Factors affecting denitrification are:

– Organic carbon: denitrifying bacteria can use different types of organic matter as carbon and energy sources. Organic matter in wastewater, the addition of methanol, ethanol, acetic acid or other organic compounds, can be used for denitrification. In biological denitrification, it is important to estimate the mass of COD required to reduce nitrates. Generally, approximately 3.5–4 g of biodegradable organic carbon are needed per gram of reduced NO_3^-. However, this ratio depends largely on the nature of the carbon and the microbial ecology. Generally speaking, since denitrifying bacteria grow with high cell yields (am), a high COD/N ratio is required.

– Temperature: the denitrification process is temperature-dependent, with the denitrification rate increasing at higher temperatures. The reported optimal temperature varies from 20 to 35°C. As the growth rate of denitrifying agents decreases with a drop in wastewater temperature, this leads to a reduction in denitrification capacity during winter. Denitrification is considerably inhibited at wastewater temperatures below 5°C.

– Oxygen: oxygen inhibits the denitrification process, which occurs under anoxic conditions. If oxygen is present, bacteria will use molecular oxygen instead of that from nitrates and nitrites as an electron acceptor. It is important to note that the actual oxygen concentration experienced by bacteria within activated sludge flocs or biofilms may be lower than the overall dissolved oxygen concentration that is measured. An overall concentration lower than 1 mg $O_2.L^{-1}$ is often recommended. However, even if a very limited amount of oxygen is introduced into the denitrification tank (DN), it is immediately consumed by the denitrifying bacteria. Denitrification can therefore still occur with a limited concentration of O_2. In an activated sludge system, bacteria group together, forming sludge flocs. Within biofilms, oxygen diffuses throughout the floc, creating an oxygen gradient. This allows denitrification to occur in the internal zones of the flocs.

– pH and alkalinity: the optimal pH for denitrification is between 7 and 9, with the denitrification rate decreasing sharply outside this range. In most cases, the pH is stable between 7 and 8 in biological systems, suggesting that pH should only slightly affect denitrification. The denitrification process produces alkalinity: approximately 0.07 meq of alkalinity per mg of reduced N-NO_3. An optimal pH is between 7 and 9.

– Nutrients: nutrient requirements are identical to those of other heterotrophic microorganisms. Phosphorus can be a limiting factor if precipitation is used prior to the biological process.

1.2.3.2. *Autotrophic denitrification*

Autotrophic denitrifying bacteria are capable of reducing oxidized forms of nitrogen, using carbon dioxide or bicarbonate.

A very wide diversity of bacterial genera are capable of denitrification; several classes of bacteria include:

– Chemolithoautotrophs and heterotrophs that use organic carbon.

– Chemolithoautotrophs, capable of oxidizing elemental sulfur (S°) and other sulfur compounds such as sulfites (SO_3^{2-}), sulfides (S^{2-}) and thiosulfate ($S_2O_3^{2-}$), which can be classed into two groups:

- Obligate chemolithoautotrophic bacteria that can only be used with sulfur, such as *Thiobacillus denitrificans*, *Thiomicrospira denitrificans* and *Sulfurimonas denitrificans*. These reactions involve denitrification processes that occur in the presence of sulfur compounds. They are practically limited to autotrophic growth, as they can only obtain energy by oxidizing sulfur compounds (Kuenen and Robertson 1992).

- Facultative chemolithoautotrophic bacteria, such as *Thiobacillus versutus*, *Thiobacillus thyasiris*, *Paracoccus denitrificans* and *Thiosphaera pantotropha*, which use sulfur as an energy source but can also grow heterotrophically by deriving their energy from an organic source. Autotrophic bacteria can oxidize an inorganic substrate such as sulfur; therefore, granular sulfur is used as a bacterial substrate. This produces energy used by microorganisms for growth. During autotrophic denitrification with elemental sulfur, the carbon source that the bacteria consume is carbon dioxide. This type of system does not require external intervention to enrich the environment with a carbon source; it is completely self-sustaining.

Autotrophic denitrification primarily results in sulfates with the production of a H^+ ion as its end product: sulfuric acid is generated, which lowers the pH (6.0 to 6.5).

The use of *Thiobacillus denitrificans* can allow the reduction of nitrates to nitrogen through a variety of sulfur compounds (S^{2-}, S, $S_2O_3^{2-}$, $S_4O_6^{2-}$, SO_3^{2-}) as they are oxidized into sulfates. Sulfur is a constituent of living matter. It makes up approximately 1% of the cell's dry weight and is a component of amino acids; for example, in the form of –SH groups (cysteine), disulfides (cystine) or thioethers (methionine). It is a vital element of coenzymes (thiamine, biotin, lipoic acid). It is present in the cartilage of higher animals (chondroitin sulfuric acid), in various natural products produced by plants, in certain antibiotics (penicillin) and in nature primarily in the form of sulfates.

Generally speaking, sulfur can be present in the form of elemental sulfur, sulfides, disulfides, sulfates, thiosulfates and even thiocyanates. Its use depends on its state. Sulfur is assimilated by animals and plants in the form of sulfates, which are extremely abundant in nature.

For example, Winogradsky (1890) showed that cell cytoplasm granules contain colloidal elemental sulfur. This sulfur, accumulated inside the bacterium, comes from the oxidation of sulfides in the environment. When cells lack sulfides, they oxidize the cytoplasmic colloidal sulfur into thiosulfates, sulfites and sulfates, which are then transformed into sulfides by other bacterial species.

As with nitrifying bacteria in the nitrogen cycle, these bacteria derive their energy from the oxidation of reduced sulfur compounds. Sulfate-reducing bacteria assimilate sulfates, transforming them into sulfides. These sulfides are then assimilated by sulfate-oxidizing bacteria, which transform them back into sulfates.

This establishes an equilibrium between the two bacterial flora, which cycle sulfur from its reduced to its oxidized state and vice versa.

1.2.3.2.1. Sulfur-oxidizing bacteria

There are many forms of sulfur-oxidizing bacteria, both morphologically and physiologically. Winogradsky showed that filamentous bacteria *Beggiatoa alba* oxidize sulfides aerobically in two steps. The first step is the oxidation of sulfides to elemental sulfur, which accumulates in the cytoplasm in the form of granules:

$$H_2S + ½\,O_2 \rightarrow S° + H_2O\ \Delta G° = 171.4\ kJ.mole^{-1}$$

When the sulfide is depleted, the elemental sulfur granules are oxidized into sulfates:

$$S° + 3/2\,O_2 + 3H_2O \rightarrow SO_4^{2-} + 2H_3O^+\ \Delta G° = -493.2\ kJ.mole^{-1}$$

These bacteria use CO_2 as the sole source of carbon. Other types of bacteria, like *Thiobacillus*, are responsible for oxidation into sulfates, sulfides, thiosulfates, elemental sulfur and thiocyanates. Various species exist depending on whether the environment is aerobic or anaerobic.

In aerobic conditions, the reactions are as follows:

– With *Thiobacillus thioxidans*:

$$4S° + 6O_2 + 12H_2O \rightarrow 4SO_4^{2-} + 8H_3O^+\ \Delta G° = -575.1\ kJ.mole^{-1}$$

– With Thiobacillus thioparus:

$$H_2S + 2O_2 + 2H_2O \rightarrow SO_4^{2-} + 2H_3O^+\ \Delta G° = -780.4\ kJ.mole^{-1}$$

– With Thiobacillus thioparus:

$$S_2O_3^{2-} + 2O_2 + 3H_2O \rightarrow 2SO_4^{2-} + 2H_3O^+\ \Delta G° = -1\,060\ kJ.mole^{-1}$$

The stoichiometries of some denitrification reactions using sulfur compounds as an energy source are given below:

$$0.844.\ S_2O_3^{2-} + NO_3^- + 0.347CO_2 + 0.086HCO_3^- + 0.086NH_4^+ + 0.434H_2O \rightarrow 1.689SO_4^{2-} + 0.500N_2 + 0.086C_5H_7O_2N + 0.697H^+$$

$$0.421H_2S + 0.421HS^- + NO_3^- + 0.346CO_2 + 0.086HCO_3^- + 0.086NH_4^+ \rightarrow 0.842SO_4^{2-} + 0.500N_2 + 0.086C_5H_7O_2N + 0.434H_2O + 0.262H^+$$

$$1.10.\ So + NO_3^- + 0.76H_2O + 0.40CO_2 + 0.08NH_4^+ \rightarrow 1.10SO_4^{2-} + 0.50N_2 + 0.08C_5H_7O_2N + 1.28H^+$$

Thiobacillus are spore-forming, rod-shaped Gram-negative bacteria, with sizes between 1 and 3 μm. Beijerinck (1910) showed that *Thiobacillus thioparus* develop at a pH close to neutral and their growth ceases around pH 5. However, *Thiobacillus thioxidans* only develop at pH 5 and their growth can lower the pH to 0.

In anaerobic conditions, anaerobic facultative *Thiobacillus* use nitrates as an oxygen source, generally *Thiobacillus denitrificans*.

The equation is as follows:

$$5S° + 6NO_3^- + 6H_2O \rightarrow 5SO_4^{2-} + 4H_3O^+ + 3N_2$$

Beijerinck (1910) showed that the optimum pH for the growth of these bacteria was between 6.2 and 7.4.

1.2.3.2.2. Sulfate-reducing bacteria

Desulfovibrio desulfuricans have often been identified as sulfur-reducing bacteria. These are non-spore-forming, Gram-negative, highly motile bacteria, between 1 and 2 μm in size, and are strict anaerobes. They reduce sulfates to sulfides using organic aromatic compounds such as dicarboxylic acids, pyruvates or lactates as an energy source.

1.3. Factors influencing nitrification

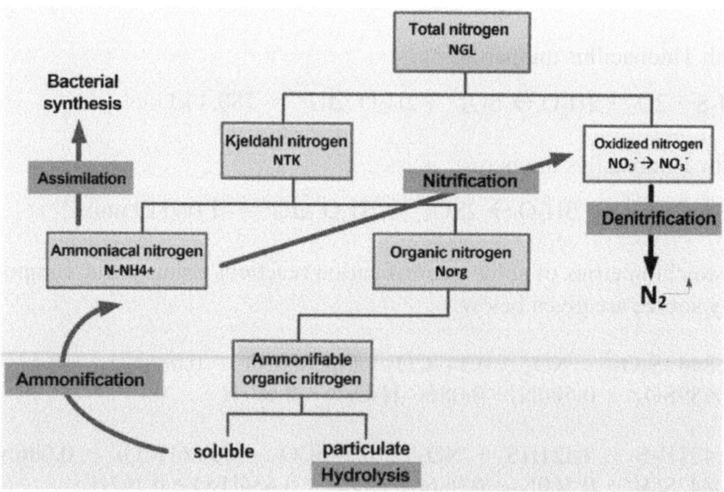

Figure 1.6. *Factors involved in nitrogen removal*

Human activity contributes to increased denitrification, among other things, through the use of fertilizers that add ammonia compounds to the soil (NH_4^+, NH_3) and nitrates (NO_3^-). The use of fossil fuels in engines or thermal power plants transforms nitrogen into nitrogen oxide NO_2^-. With N_2 and CO_2, denitrification releases a small amount of nitrogen oxide N_2O into the atmosphere. The concentration of this gas is low, 300 ppb (parts per billion). However, it should be noted that a molecule of N_2O is 200 times more effective than a molecule of CO_2 in creating a greenhouse effect. It is currently estimated that the concentration of atmospheric N_2O increases annually by 0.3%, and this increase is almost entirely linked to emissions due to soil denitrification. Studies of Antarctic ice cores have shown that the concentration of atmospheric N_2O was 270 ppb at the end of the last ice age (10,000 years ago). This concentration remained at that level until the industrial era, when it jumped to its current level of 300 ppb, an increase of 11%. For the nitrification process to be efficient, the concentration of organic matter must be low. Otherwise, fast-growing heterotrophic microorganisms will consume the oxygen necessary for their survival. Factors affecting the nitrification process include temperature, oxygen concentration, pH, alkalinity and inhibitory substances in wastewater.

1.3.1. *Temperature*

Nitrification is strongly affected by temperature; temperature alters the rate of synthesis of nitrifying bacteria below 30°C. Between 30 and 35°C, growth is little influenced by temperature, and it then becomes impossible beyond 35°C (see Figure 1.7).

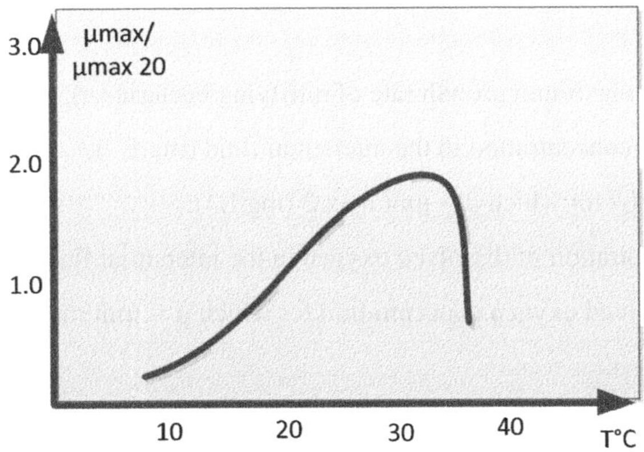

Figure 1.7. *Influence of temperature on the maximum growth rate of autotrophs (Henze et al. 1995)*

A drop in temperature from 16°C to 6°C will reduce the nitrification rate by 2.5 times. Temperature is an important parameter in all biological reactions and has a pronounced effect on the growth rate of nitrifying organisms.

1.3.2. *Growth rate*

The growth rate of autotrophic bacteria is given in Table 1.3.

Temperature (°C)	Growth rate	
	Maximum (μmax) (d^{-1})	**Observed on site (d^{-1})**
20	0.7	0.47
10	0.21–0.31	0.14–0.21

Table 1.3. *Growth rate of autotrophic bacteria*

COMMENTS ON TABLE 1.3.– *Maximum (μmax): obtained in the laboratory without limiting factors; observed on site: function of the maximum growth rate and a number of factors such as the concentration of N-NH_4^+, oxygen concentration, temperature and minimum oxygen presence.*

$$\mu\text{nit site} = \mu\text{nit max}\frac{NH_4}{K_{NH4}+NH_4}\,\frac{O_2}{KO_2+O_2}$$

where:

– μnit max = maximum growth rate of nitrifying bacteria (/t);

– NH_4 = NH_4 concentration in the interstitial fluid ($mg.L^{-1}$);

– K_{NH4} = $[NH_4]$ for which μ = μnit max/2 ($mg.L^{-1}$);

– O_2 = concentration of dissolved oxygen in the interstitial fluid ($mg.L^{-1}$);

– K_{O2} = dissolved oxygen concentration for which μ = μnit max/2 ($mg.L^{-1}$).

Furthermore:

$$\mu\text{nit max (T)} = \mu\text{nit max (20°C)} \times \theta^{(T-20)}$$

With θ = 1.06, we obtain:

$$\mu\text{nit max (20°C)} = 0.8\ d^{-1}$$

At 10°C:

$$\mu nit\ max\ (T) = \mu nit\ max\ (20°C) \times \theta^{(T-20)}$$

$$\mu nit\ max\ (10°C) = 0.8 \times 1.06^{-10} = 0.45\ d^{-1}$$

The temperature change constant (θ) takes on quite different values depending on the author: 1.02 (Oleszkiewicz and Berquist 1988), 1.024 (Argaman 1994; Mines and Sherrard 1999), 1.044 (Funamizu and Takakuwa 1994), 1.06 (Deronzier et al. 2002), 1.103 (Lesouef 1992) and 1.165 (McCartney and Oleszkiewicz 1990).

Henze et al. (1995) propose the following:

$$\mu n\ max(pH) = \mu n\ max(pH\ optimal)\frac{KpH}{KpH + 10^{(pHoptimal-pH)}}$$

with KpH = 200.

In winter, complete nitrification depends on a combination of several factors such as the volume of the aeration tank, suspended solids (SS) and pH, whereas in warmer periods, lower sludge levels and a less stringent pH value are sufficient.

1.3.3. *Oxygen*

Nitrifying bacteria are more sensitive to low oxygen concentrations than heterotrophic microorganisms. At O_2 concentrations below 2–3 mg.L^{-1}, this will likely be a limiting factor depending on the type of biological treatment used.

1.3.4. *pH and alkalinity*

The nitrification process is highly dependent on pH and alkalinity.

The optimal pH values for *Nitrosomonas* are located between 8 and 9, while at pH 7, their activity decreases by 50%. For *Nitrobacter*, the maximum growth rate occurs at pH 7–8.6.

The pH of the wastewater will determine the ammonia concentration (NH_3) and nitric acid (HNO_2), which in turn can inhibit nitrifying bacteria. The nitrification process also consumes alkalinity, which can lead to a reduction in pH.

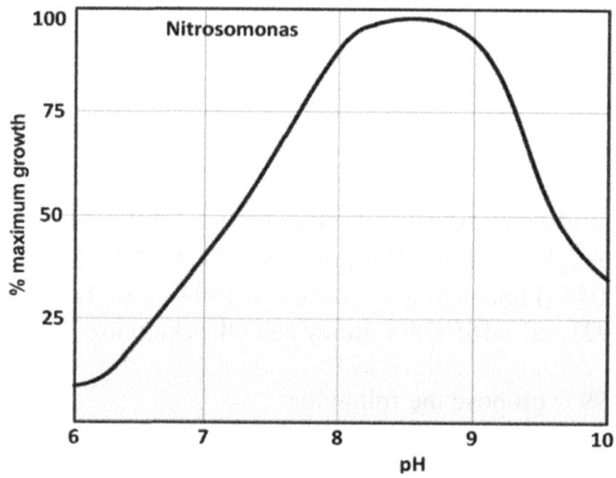

Figure 1.8. *Influence of pH on the growth of* Nitrosomonas *at 20°C (Wild 1994)*

1.3.5. *Nutrients*

In municipal wastewater, the amount of nutrients is normally sufficient for microbial growth. However, phosphorus can become a limiting factor when pre-precipitation is used upstream of the biological process. At ortho-P concentrations < 0.15–0.2 mg.L^{-1}, the nitrification reaction could be inhibited.

1.3.6. *Dissolved oxygen and nitrogen substrate*

Dissolved oxygen and ammoniacal nitrogen are the essential reactants for nitrification. When present in sufficient quantities, the growth of autotrophic bacteria is maximized. When one of the two reactants becomes limiting, bacteria have more difficulty accessing it, which slows their growth rate. Manufacturers typically describe the influence of dissolved oxygen and nitrogen substrate on the growth rate of autotrophs using a double Monod equation:

$$\mu\text{aut}(T) = \mu\text{aut}\max(T)\frac{O_2}{O_2 + KO,\text{aut}}\frac{N - NH_4}{N - NH_4 + K_{NH4}}$$

Practical and easy to implement, this type of relationship is also used to predict the growth rate of heterotrophs during assimilation. In an oxygen-limited environment, autotrophic and heterotrophic biomasses compete. The course of the reactions depends on the bacteria's ability to access oxygen at low concentrations. Specific to each biomass, this affinity is expressed as a half-saturation constant.

Heterotrophic bacteria, which more readily use oxygen at low concentrations, are characterized by a lower half-saturation constant (KO, heterotrophs = 0.2 $mgO_2.L^{-1}$) than that of autotrophs (KO, autotrophs = 0.2 $mgO_2.L^{-1}$). The value of K_{NH4} is equal to 0.1 $mg.L^{-1}$ (Henze et al. 1986).

Thus, the main factors influencing the growth of autotrophic bacteria in activated sludge are pH and temperature, which affect the internal mechanisms of the bacteria. They determine the maximum growth rate of nitrifying biomass (µA, maxi), ammoniacal nitrogen and dissolved oxygen through their availability. Their concentrations modulate the maximum growth rate of autotrophic biomass.

Net growth, resulting from growth and death, taking into account both the limitation of the dissolved oxygen concentration and the nitrogen substrate, is then defined for a continuous 24 h oxygenation period by the equation:

$$\mu aut = \mu aut\, max(20°C)\, \Theta^{T-20} \frac{O_2}{O_2 + KO, aut} \frac{N - NH_4}{N - NH_4 + K_{NH4}} - baut$$

i.e.:

$$\mathrm{\mu aut} = \mathrm{\mu aut\,max(20°C)}\, \Theta^{\mathrm{T}-20} \frac{O_2}{\mathrm{O_2 + 0.4}} \frac{\mathrm{N - NH_4}}{\mathrm{N - NH_4 + 0.1}} - \mathrm{baut}$$

If the oxygenation time is not continuous, this parameter is integrated and the equation becomes:

$$\mathrm{\mu aut} = \mathrm{\mu aut\,max(20°C)}\, \Theta^{\mathrm{T}-20} \frac{\mathrm{O2}}{\mathrm{O2 + 0.2}} \frac{\mathrm{N - NH_4}}{\mathrm{N - NH_4 + 0.1}} \frac{\mathrm{t}}{24} - \mathrm{baut}$$

where t is the operating time of the aerators (h).

Table 1.4 summarizes the main growth parameters for the nitrogen treatment simulation.

Reference 10°C		Symbol	Units	Values
Maximum growth rate		µ max	d^{-1}	0.45
Saturation constant	**Substrate**	Ks NH_4	g $N\text{-}NH_4^+.m^{-3}$	0.1
	Oxygen	Ks O_2	g $O_2.m^{-3}$	0.2
Cellular yield of autotrophs		Y max	g $MVS.g^{-1}$ $N\text{-}NO_3$ formed	0.24
Endogenous respiration constant		b autotrophs	d^{-1}	0.13
Temperature constant for µ max		Θ	$°C^{-1}$	1.06

Table 1.4. *Growth parameters for autotrophic bacteria*

1.4. Nitrification in an activated sludge system

The implementation of nitrification in activated sludge depends on several factors that influence the sizing of the biological reactors.

1.4.1. *Influence of mass loading*

The mass loading corresponds to the quantity of BOD_5 applied daily to the biomass present in the system (kg of MVS). Mass loading has several consequences: on the plant's efficiency, on the sludge's volatile solids (VSS) content, on the culture age, on the specific sludge production and on oxygen consumption. As the mass loading (Cm) increases, the age of the sludge decreases. Indeed, the increase in organic matter leads to a significant increase in the sludge mass, and therefore in the mass to be extracted (Xa). The extractions gradually remove the autotrophic bacteria from the environment.

The classification of processes is established according to Table 1.5.

Achieving the classically accepted rejection level (COD < 125 $mg.L^{-1}$, BOD_5 < 25 $mg.L^{-1}$ and TSS < 35 $mg.L^{-1}$) requires operation at the mass loads shown in Table 1.6.

Mass load (kg BOD.kg.MVS^{-1}. d^{1})	Process
Cm < 0.1	Extended ventilation
0.1. < Cm < 0.2	Low load
0.2 < Cm < 0.5	Medium load
Cm > 0.5	High load
Cm > 1	Very high load

Table 1.5. *Classification of processes based on mass loading*

Mass load (kg BOD.kg.MVS^{-1}. d^{-1})	Temperature	Process
Cm ≤ 0.35	10°C	Medium load
Cm ≤ 0.8	20°C	Heavy load

Table 1.6. *Influence of temperature for constant NH_4 yield*

The influence of mass loading on nitrification is as follows:

– A high proportion of organic matter promotes the growth of heterotrophic bacteria.

– The cellular yield of autotrophs (*Nitrosomonas* and *Nitrobacter*) is much lower than that of heterotrophs. The autotrophic population declines rapidly due to sludge removal.

– For optimal nitrification to occur, it is necessary to establish a balance between heterotrophs and autotrophs; the decomposed organic matter will form CO_2 which will be used by autotrophs to synthesize new cells.

– The increase in mass loading causes oxygen deficiencies in the tank and affects nitrification.

Mass loadings less than 0.03–0.04 kg of $BOD_5.kg^{-1}$ MVS. d^{-1} are too low. This leads to the formation of small flocs due to slight deflocculation linked to a limited amount of substrate. These flocs result in lower sedimentation rates, which can negatively impact discharge during hydraulic surges in the clarifier. High mass loadings, in addition to the non-compliant yield (yield decreases with increasing mass loading), also lead to the formation of small flocs. Furthermore, sudden changes in mass loading (from a low to a high value) frequently result in the growth of filamentous bacteria. The mass loading remains fairly stable if the sludge content is kept constant by the operator (Canler et al. 2007).

1.4.2. *Influence of TSS/MVS concentration in the tank*

The concentration of SS has consequences for the sizing of the biological reactor. For the average load, an SS concentration of up to 3 g $MVS.L^{-1}$ is used, that is, a TSS rate of 4 $g.L^{-1}$. An excessively high SS level in the biological reactor, combined with typical recirculation rates of 150%, can cause system blockages due to limited thickening resulting from a degradation of the sludge index and trigger sludge removal (Canler et al. 2007). In the case of low loading rates or prolonged aeration, increasing the concentration in the aeration tank is more advantageous. It is possible to work with higher concentrations (4.5 $g.L^{-1}$).

1.4.3. *Influence of loading rate*

The volumetric charge or loading rate of the system is obtained from the ratio of the quantity of BOD_5 to be treated per day and the reactor volume or from the product of the Cm and the MVS concentration:

Loading rate (Cv) = mass load x MVS concentration

i.e.:

– Cv = flow rate BOD_5 (kg. d^{-1})/reactor volume (m^3);

– T retention time (hour or day) = reactor volume (m^3)/hourly or daily flow rate;

– Cv = BOD_5 concentration × 1/retention rate.

This volumetric loading rate therefore corresponds to the retention time of the wastewater to be treated with biomass in the biological reactor. We obtain a Cv value of less than 0.6 kg $BOD_5.m^{-3}$. d^{-1}.

1.4.4. *Influence of sludge age*

The sludge age is the ratio (G) of the mass of sludge present in the aerator (Xa) to the mass (Xb) of sludge removed daily. It is expressed in days, that is, G = Xa/Xb.

This quotient therefore represents the average retention time of the sludge in the active part of the system and is directly related to its state of evolution. Achieving a high sludge age to maintain autotrophic biomass requires working at a specific mass loading. The maximum mass loadings and minimum sludge ages required for the nitrification process as a function of temperature are presented in Table 1.7.

Temperature in the biological reactor	10°C	15°C	20°C
Mass load (kg of $BOD_5.kg^{-1}$ of MVS. d^{-1})	0.1	0.15	0.2
Age of the sludge (days)	16–18	10–12	8–9

Table 1.7. *Relationships between temperature, mass loading and age of the sludge*

1.4.5. *Nitrification rate*

$$Vnit\ T = A\ Cv_N\ \theta^{(T-20)}$$

where:

– A constant = 0.1161 at 20°C;

– Vnit T = nitrification rate at temperature T (mg $N.L^{-1}.h^{-1}$);

– T = temperature (°C);

– θ = corrective coefficient = 1.06;

– C_{VN} = NTK volumetric load (g NTK input.m^{-3} in reactor. d^{-1}).

1.4.6. *O_2 requirement*

Under average load as well as during prolonged ventilation, it is necessary to cover the sum of the needs corresponding to:

– the oxidation of organic matter;

– the auto-oxidation of sludge (endogenous respiration);

– nitrogen oxidation (reduced by the oxygen release observed during denitrification).

Furthermore, the daily operating time of the aeration system must allow for periods of anoxia necessary for denitrification. These periods correspond to the difference in the amount of oxygen required to oxidize ammonia nitrogen to nitric nitrogen (nitrification) and the amount of oxygen recovered by the reduction of nitric nitrogen to nitrogen gas (denitrification), which can be written as:

$$NH_4OH + 2O_2 \rightarrow NO_3H + 2H_2O$$

– Consumption of 4.57 mg O_2.mg^{-1} N oxidized:

$$2NO_3H \rightarrow N_2 + H_2O + 5/2O_2$$

– Recovery of 2.86 mg O_2. mg^{-1} N reduced.

– Nitrification: oxygen requirements due to nitrification:

$$O_2n = 4.57 \times \text{N nitrified}$$

– Denitrification: the amount of oxygen recovered by denitrification:

$$O_2\,dn = 2.86 \times \text{N denitrified}$$

1.4.6.1. *Nitrogen balance*

1.4.6.1.1. *Nitrogen for synthesis*

A quantity of nitrogen (in the form of TNK) is removed simultaneously with BOD and is quantitatively present in the sludge produced. The concentration of synthetic nitrogen is an estimated 5% of the removed BOD flow:

N synthesis (Nsy) = 5% BOD removed

1.4.6.1.2. Nitrogen removed by the nitrification–denitrification process

If nitrogen removal by synthesis alone is insufficient to meet the required nitrogen standard for the treated effluent, additional nitrogen must be removed by nitrification–denitrification. The calculation is performed retrospectively as follows:

Nitrified nitrogen (Nnit) = NTK influent – NTK effluent – Nsy (kg.d^{-1})

Denitrified nitrogen (Ndenit) = N nitrified – N.NO_3 effluent (kg. d^{-1})

These values are established by taking nitrogen standard values for NTK effluent and N-NO_3 effluent.

For example, for 10 mg.L^{-1} NGL, NTK = 5 mg.L^{-}1 and N-NO_3 = 5 mg.L^{-1}. For 20 mg.L^{-1} NGL, NTK = 10 mg.L^{-}1 and N-NO_3 = 10 mg.L^{-1}.

The nitrification objective is then considered to be achieved provided that a mass of MVS is maintained in aeration (in the aerobic zone) such that: nitrified nitrogen/aerobic X $\leq$ 3 mg.g^{-1} MVS.h^{-1}.

1.4.6.2. *Minimum sludge age required for nitrification*

The minimum sludge age required for nitrification is calculated based on the aerated volume and is a function of the temperature (T): this formula is valid for a treatment plant serving a population greater than 100,000 individuals:

Gnit = FS x 2.13 $\times$ 1.103(15-T) in days

FS = 2.3 for a plant treating a population greater than 100,000 individuals.

The minimum sludge age recommended for nitrification is typically around 8 days.

Oxygen requirements can also be defined in relation to the entering BOD_5. The total oxygen demand expressed relative to the BOD_5 applied is calculated from:

– Specific oxygen requirements due to the elimination of BOD_5 (O_{RBOD5}):

O_{RBOD5} (kg O_2.kg^{-1} BOD_5)

– Specific oxygen requirements due to nitrogen removal (O_{RN}):

O_{RN} (kg O_2.kg^{-1} BOD_5)

The specific needs (both for BOD_5 as with nitrogen) are defined in relation to the entering BOD_5:

$$O_2\ (kg\ O_2.d^{-1}) = Q\ (O_{RBOD5} + O_{RN})\ Lo/1{,}000$$

where Lo is the BOD_5 wastewater entering the biological reactor ($mg.L^{-1}$).

The assessment of specific needs for the BOD_5 and nitrogen is given by:

– The specific oxygen consumed by the BOD_5 (O_{RBOD5}): the oxygen consumed by carbon pollution depends on the age of the sludge and the temperature:

$$ORBOD5 = \frac{0.144\ G\ f}{1 + 0.08G\ f} + 0.5$$

where O_{RBOD5} is in kg $O_2.kg^{-1}$ BOD_5, G is the sludge age (d) and f is the temperature correction factor.

– The oxygen demand specific to nitrification, taking into account the oxygen recovered by denitrification O_{RN}:

$$ORN = \frac{Q(4.6\ N - NO3 - 1.7\ N + NO3\ denitrified)}{Q\ x\ BOD5}$$

where O_{RN} is in kg $O_2.kg^{-1}$; BOD_5 and $N\text{-}NO_3$ are in $kg.d^{-1}$, and Q is in $m^3.d^{-1}$.

1.4.7. *Dynamic model of nitrification*

If we assume complete mixing, that is, instantaneous mixing of the inlet water throughout the entire volume of the aeration tank, we can describe the nitrification process by performing mass balances on both the nitrogen removed and the sludge present. Consider an activated sludge treatment plant (see Figure 1.9) consisting of a biological reactor with volume V containing sludge with concentration Xa receiving an effluent of concentration Lo at a daily volume Q. To return the sludge from the settling tank to the reactor inlet, a recycling rate r is applied.

Part of the removed carbon pollution is transformed into biomass, most of which is extracted (concentration Xb with a flow rate σQ). The treated effluent obtained at the outlet has a soluble pollution (Lf) and SS (Xf) concentration. X denotes MES; X_V denotes MVS; X_N is the nitrifying mass; Xh denotes heterotrophic organisms; b is the endogenous coefficient; μ is the growth rate; am is the cell yield; t is the retention time; N_H is the ammoniacal nitrogen; N_N is the nitrate nitrogen; and NKT is the Kjeldahl total nitrogen (ammoniacal and organic nitrogen).

When the system operates in steady state, the extracted sludge allows the average retention time of the sludge in the system (also called "sludge age" and denoted SRT) to be defined as the total mass of sludge divided by the quantity of sludge extracted daily.

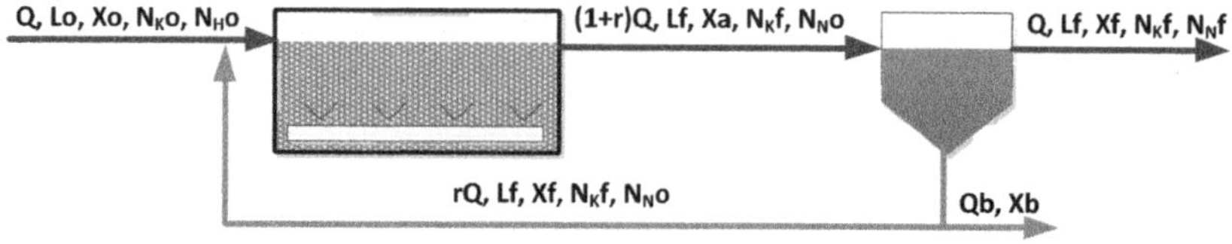

Figure 1.9. *Inlet and outlet characteristics of an activated sludge plant*

The balance sheet for organic matter is written:

$$V\,dL = Q\,Lo\,dt + rQ\,Lf\,dt - (1+r)\,Q\,Lf\,dt - (V\,\mu\,Xv\,dt)/am$$

In steady state, dL/dt = 0.

By setting the hydraulic retention time, t = V/Q, it becomes:

$$\frac{Lo}{t} + \frac{rLf}{t} - \frac{Lf}{t} - \frac{rLf}{t} - \frac{\mu Xa}{am} = 0$$

After simplification, we obtain:

$$\frac{Lo - Lf}{t} = \frac{Q(Lo - Lf)}{V} = \frac{\mu Xa}{am}$$

Hence:

$$\mu = \frac{am(Lo - Lf)}{Xa\,t} = \frac{\frac{amLo(Lo - Lf)}{lo}}{Xa\,t}$$

Consider the parameters mass charge (Cm) and carbon removal efficiency (ρ) defined by the relationships:

Cm = Lo/Xt

This is the daily BOD applied flux relative to the amount of active biomass present in the aeration tank.

ρ = (Lo-Lf)/Lo represents the purification efficiency.

The biomass growth rate is then expressed by the equation:

$$\mu = \text{am Cm}.\rho$$

Furthermore, the age of the G (or SRT) sludge is written as:

$$G = \frac{V\,Xa}{(1-\sigma)Q\,Xf + \sigma\,QXb} = \frac{Xa\;t}{(1-\sigma)Xf + \sigma Xb}$$

Assuming biomass losses (Xf) in the outlet water are negligible, the sludge age equation simplifies to:

$$G = \frac{Xa\;t}{\sigma Xb}$$

where rX is the rate of biomass growth. The biomass balance is written as:

$$rX = (\mu - b)V\,Xa = (1-\sigma)Q\,Xf + \sigma\,QXb = VXa/G$$

$$(\mu - b) = 1/G = \text{am.Cm}.\rho$$

1.4.7.1. *Empirical relations*

Based on the results obtained from real-world installations, wastewater treatment plant designers have determined the influence of temperature on the maximum autotrophic growth rate (Martin 1979):

$$\mu \text{ autotrophs, maxi} = 0.13 * 1.09^{(T-15)} \text{ (expressed in } d^{-1}\text{)}$$

Assuming that the ammoniacal nitrogen concentration in the aeration tank can be estimated using the average concentration at the plant outlet, the minimum retention time required to maintain nitrification is expressed by the following equation:

$$G \text{ mini} = 1/(\mu aut - baut)$$

$$G\text{ mini} = \frac{1}{\mu aut - baut} = \frac{1}{\mu\text{ aut, mini}\,\frac{(N-NH4)f}{(N-NH4)f + 1.5} - baut}$$

with:

- μaut: autotrophic growth rate;
- baut: endogenous respiration coefficient of autotrophs.

Neglecting baut, the equation finally becomes:

$$G\ \text{mini} = \frac{(N - NH4)f + 1.5}{(N - NH4)f} \frac{1}{0.13\ 1.09^{(T-15)}}$$

Another approach, for nitrification, involves defining a fictitious sludge age based solely on the sludge mass in aerated areas:

G aerated = X aerobic/sludge produced

the minimum value per G aerated is then generally taken as a function of the temperature:

G aerated mini = 4.5 x 0.914(T – 20)

The denitrification objective is achieved provided that a mass of MVS is maintained in the anoxic zone, such as: denitrified nitrogen/Xv anoxic < 2–4 $mg.g^{-1}$ $MVS.h^{-1}$.

The total mass of MVS to be maintained is then:

Xt = X aerobic + X anoxic

No excess sludge is produced for nitrogen removal.

1.5. Denitrification in activated sludge and denitrification rate

The implementation of nitrification in activated sludge depends on several factors that influence the sizing of the biological reactors.

The kinetics of the denitrification reaction with activated sludge can be expressed by:

$v_{DN} = dN\text{-}NO_3/dt = KX$

with:

– $dN\text{-}NO_3/dt$ = denitrification rate (g $N\text{-}NO_3.L^{-1}.h^{-1}$);

– $N\text{-}NO_3$ = nitrate concentration (mg $N\text{-}NO_3.L^{-1}$);

– t = time (hours);

– X = biomass concentration (mg $MVS.L^{-1}$);

– K = specific denitrification constant (mg $N\text{-}NO_3.mgMVS.h^{-1}$).

This expression indicates that the nitrate–time relationship is linear and independent of the nitrate concentration; that is, the rate is of zero order with respect to the nitrate concentration and is a function solely of the sludge concentration.

Since the denitrification reaction is controlled by active biomass, we write, for simplicity:

$$\Delta \text{N-NO}_3 = \text{K Xa } \Delta t$$

The denitrification rate is given by the following relationship, correlated with temperature according to Arrhenius law:

$$v_{DN}\ (T) = v_{DN}\ (20°C)\ \Theta^{\ (T-10)}$$

However, as the kinetics of denitrification can be disrupted by the presence of dissolved O_2, this inhibition is taken into account. The formula becomes:

$$vDN = \frac{1-1.42\ Y}{2.86}\ \frac{K\,S\,X}{Ks+S}\ \frac{N-NO3}{Kn+N-NO3}\ \frac{K'O}{K'O+O2}\ \eta$$

with:

– v_{DN}: denitrification rate (mgN-NO_3^-.L^{-1} h^{-1});

– Y: cell synthesis efficiency of heterotrophs (mg MVS.mg^{-1} substrate consumed);

– K: maximum specific substrate consumption rate (h^{-1});

– X: biomass concentration in the reactor (mg MVS L^{-1});

– S: concentration of the biodegradable soluble substrate (mg L^{-1});

– KS: constant half-rate of substrate consumption reaction (mg L^{-1});

– N-NO_3: nitric nitrogen concentration (mgN-NO_3^-.h^{-1});

– KN: half-rate constant of nitrate reaction (mgN-NO_3^-.h^{-1});

– K'O: constant relative to the inhibition of O_2 for the reduction of nitrates (mgO_2.$^{-1}$);

– η: fraction of heterotrophic bacteria using nitrates as an O_2 source.

v_{DN} at 20°C is the denitrification rate at 20°C in the DN, generally equal to 3 mg N-NO_3 gMVS^{-1}.h^{-1}.

In activated sludge tanks, the vDN value often reported is 2.9–3.0 gN-NO_3^-.h^{-1} kgMVS^{-1} (Raboni 2014). Manufacturers use various empirical or semi-empirical relationships involving the specific rate of denitrification:

$$\text{V specific DN}_{20°C} = 0.029 + 0.03\ \text{Cm}$$

$$\text{V specific DN (g N-NO}_3\text{.g MVS}^{-1}\text{.d}^{-1}) = \text{Q N-NO}_3\text{/V.X}$$

Sludge production in anoxic-aerobic systems does not differ significantly from that of a comparable aerobic system, provided that the anoxic volume fraction is not too large.

K is a constant of the denitrification rate.

In an anoxic tank, the specific rate of denitrification represents the rate of nitrate reduction relative to the biomass concentration.

We can write:

$$\Delta \text{N-NO}_3 \text{ eliminated} = \text{Vanox vs Xa}$$

where $N\text{-}NO_3$ is expressed in g $N\text{-}NO_3.h^{-1}$; Vanox is the volume of the anoxic tank (m^3); vs is the specific rate of denitrification in g $N\text{-}NO_3.SS^{-1}.h^{-1}$; and Xa is the concentration of suspended sludge (SS) in the tank ($g.m^{-3}$).

The observed values are highly variable across the different sites where they are measured. It is proposed that vs can be determined from the relationship:

$$\text{vs} = 0.03 \text{ Cm} + 0.029$$

where Cm (kg $BOD_5.kg^{-1}$ SS^{-1}. d^{-1}) = Q Lo/Vanox Xanox

where Lo is the BOD of the influent ($mg.L^{-1}$) and Xanox is the sludge concentration in the anoxic tank ($mg.L^{-1}$).

We can then deduce:

$$\text{vs} = 0.03 \text{ Cm} + 0.029$$

The various parameters related to denitrification kinetics are shown in Table 1.8.

Parameter	Units	Values
Cellular yield am	g $MVS.g^{-1}$ COD	0.40
Mortality coefficient b	g $MVS.g^{-1}$ TSS d^{-1}	0.07–0.10
Maximum specific growth rate µm	g $MVS.g^{-1}$ MVS. d^{-1}	3.2
Half-speed Ks	$g.m^{-3}$	9.0
Biomass COD	g $COD.g^{-1}$ MVS	1.42
Denitrifying bacteria fraction η	g $MVS.g^{-1}$ MVS	0.50

Table 1.8. *Parameters related to denitrification kinetics*

The relationship of vs with temperature is:

$$vs_{(T)} = vs_{20}\,\Theta^{T-20}$$

Θ often has a value of 1.026.

In a denitrification configuration with an upstream anoxic tank, sludge and nitrified effluent recycling is integrated.

For a recirculation rate (R) of 2:

$$\text{vs adjusted} = \text{vso} - 0.0166\ \text{Ln Cm} - 0.0078$$

where vso is the specific velocity for an internal ratio (IR) equal to 1.

For a ratio of 3–4, the relationship becomes:

$$\text{vs adjusted} = \text{vso} - 0.029\ \text{Ln Cm} - 0.012$$

Adjusted Cm is the mass load applied to the anoxic tank ($g.g^{-1}.\ d^{-1}$).

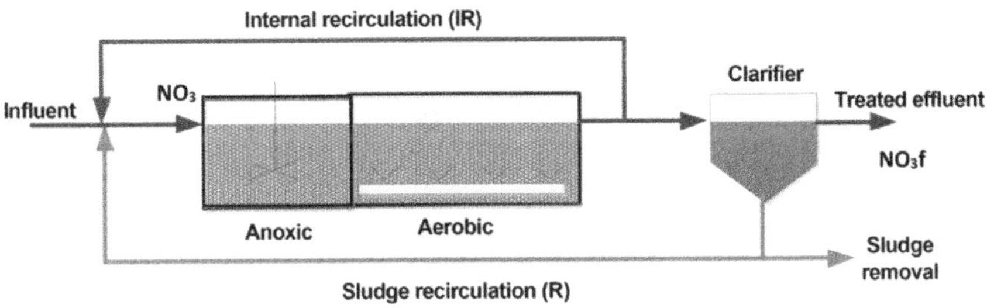

Figure 1.10. *Denitrification diagram with an anoxic tank installed upstream*

The active biomass in the anoxic tank is:

$$X\ \text{anox} = \frac{GQ}{V}\,\frac{\text{am}(\text{Lo} - \text{Le})}{1 + b\,G}$$

In the case of alternating aeration in biological tanks, as with the Biodenitro™ process (see Figure 1.11), the specific rate vs is given by:

$$\text{vs denitrification} = 0.1754\ \text{An}/(\text{am G})$$

where An is the coefficient related to the consumption of O_2 ($gO_2.g\ COD^{-1}$ removed).

The number 0.1754 is based on 2.86 g O_2 per g NO_3 and the assumption that only 50% of heterotrophic biomass is available for denitrification.

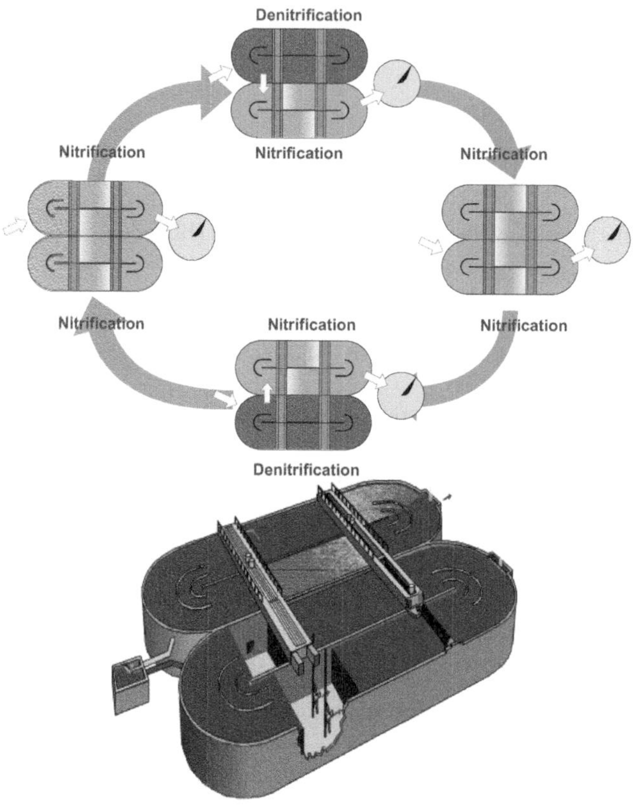

Figure 1.11. *Alternating aeration in oxidation channels: Biodenitro™ process (Veolia)*

Reducing nitrogen levels is important when effluents are discharged into sensitive water bodies that are subject to eutrophication.

1.5.1. *Specific anoxic zone for denitrification*

The total tank volume is estimated based on the desired volumetric load. This volume must then be distributed between the aerobic and anoxic zones.

Criteria:

– Mass charge in the anoxic tank Cm = 20 to 50 g $N.kg^{-1}$ $MVS.d^{-1}$.

– Volume of the anoxic tank V_{anoxic} = (25 to 40%) $V_{aerated+anoxic}$.

The volume of the anoxic tank varies depending on the nitrate flow recirculated from the clarifier and the flow supplied by the influencing material. The lower this flow, the smaller the tank.

1.5.2. *Estimation of excess sludge production*

The volume of the biological reactor (aerobic + anoxic) is defined based on the sludge age and the required sludge concentration (Xa). The Xa concentration is determined according to the treatment process and allows for the sizing of the clarifier, taking into account the Mohlman index and the most demanding hydraulic regime.

The total mass of sludge, combined with the degradation kinetics of carbonaceous and nitrogenous pollution, determines the tank volume based on the untreated water flow and the SS concentrations at the outlet. The clarifier's performance depends on various parameters: sludge mass flow, sludge thickening capacity, maximum sludge retention time and the structure's geometry.

The parameters influencing the production of biological sludge are the ratio TSS/ BOD_5, the mass load and the MVS/TSS ratio in untreated wastewater.

Sludge production can be obtained using the following relationship:

$$\text{Xb excess} = \Delta X = \text{Xmin} + \text{Xdur} + (0.83 + 0.2 \log \text{Cm})\ BOD_5 + K'\ \text{nitrified} - \text{Xs}$$

with:

– Xdur: non-biodegradable fraction of MVS = 0.15 to 0.3 (MVS) untreated water;

– Xmin: mineral fraction of TSS = TSS x (100% of MV in TSS)/100;

– Cm: applied mass load kg $BOD_5.kg^{-1}MVS.\ d^{-1}$;

– BOD_5: concentration BOD_5 in the supply water;

– K′ = 0.17 kg $nitrifying.kg^{-1}$ N nitrified;

– Xs: SS concentration in the outlet effluent ($mg.L^{-1}$);

– ΔX: excess sludge production requiring daily extraction ($kg.\ d^{-1}$).

In the case of biological phosphorus removal, the production of physicochemical sludge is taken into account.

1.5.3. *Carbon addition*

Denitrification is a carbon-consuming biological reaction. The denitrification capacity will therefore depend on the level of assimilable COD in the water being treated.

The requirement for assimilable COD for denitrification is approximately 4.6 kg $COD.kg^{-1}$ N, where N is the mass of $N.NO_3$ to be denitrified. This figure incorporates the needs for catabolism and anabolism. If the water to be treated does not contain enough assimilable carbon pollution, COD will need to be injected. The COD source that usually offers the best ratio of sludge quality produced to price is methanol.

Characteristics of methanol:

– Density = 0.79 $kg.L^{-1}$.

– COD = 1.5 $g.g^{-1}$ of methanol.

– BOD = 1 $g.g^{-1}$ of methanol.

– COD/N applied = 4–7.

– Kinetics = 2–3.5 g $N.NO_3.kg^{-1}$ MV/h.

– Sludge produced = 0.14 kg $TSS.kg^{-1}$ of methanol.

Methanol requires a specific biomass; maximum kinetics can only be reached after one month of injection.

The amount of carbon to inject is estimated by analyzing the remaining $N.NO_3$ in the treated water ($N.NO_{3outlet}$). We then calculate the flow of $N.NO_3$ that still needs to be treated by methanol injection:

– Flow $N.NO_{3\ \text{to be treated}} = (N.NO_{3inlet} - N.NO_{3\text{desired outlet}})Q$;

– Q = water flow rate to be treated in $m^3.d^{-1}$;

– COD flow to be injected ($kg.d^{-1}$) = flow $N.NO_{3\ \text{to be treated}} \times 4.6$;

– Methanol flow to be injected (kg/d^{-1}) = COD flow to inject/1.5;

– Methanol injection rate ($l.d^{-1}$) = methanol flow/0.79.

Carbon injection must take place in or just upstream of the anoxic tank (where denitrification occurs), in a well-agitated area. The flow rate of the carbon substrate is controlled by the flow rate of the water to be treated; it must be injected in slight excess. Carbon injection must be carried out at a constant volume (flow rate and duration) with sequential operation of the supply pumps. The pumps must be activated by the passage of a fixed volume of water to be treated. The pumps must operate efficiently to meet the reagent requirements for this volume of water, assuming a constant nitrate concentration. Regulation based on nitrate concentration measurements is possible.

1.6. Nitrification and denitrification processes in activated sludge

Activated sludge processes can be classified into two categories depending on whether the denitrification zone is installed upstream or downstream of the aeration zone.

1.6.1. *Pre-denitrification*

Wastewater and activated sludge come into contact and mix in an anoxic zone located upstream of an aeration zone. Nitrates formed in the aerobic zone by nitrification are recycled to the pre-DN. This process allows the use of the available organic carbon source in untreated wastewater. The denitrification yield is affected by the organic carbon concentration (BOD, COD) as well as other parameters such as biomass concentration (MVS) and temperature.

Since nitrification occurs after denitrification, the nitrates resulting from the oxidation of NH_4 in the aeration tank are ready to be discharged with the treated effluent if recycling does not occur.

Such a process typically produces an effluent with high concentrations of total nitrogen (NGL) below 20 mg N. L^{-1}.

1.6.2. *Post-denitrification*

This configuration includes an aerobic biological tank at the head of the line followed by an anoxic (non-mechanically aerated) tank. Self-generated endogenous organic matter and/or external carbon sources are used as carbon sources, because organic carbon (COD, BOD) is consumed during the first stage, with an energy cost associated with aeration. Post-denitrification can lead to the release of ammonium in the anoxic zone if only self-generated endogenous organic matter is used as carbon.

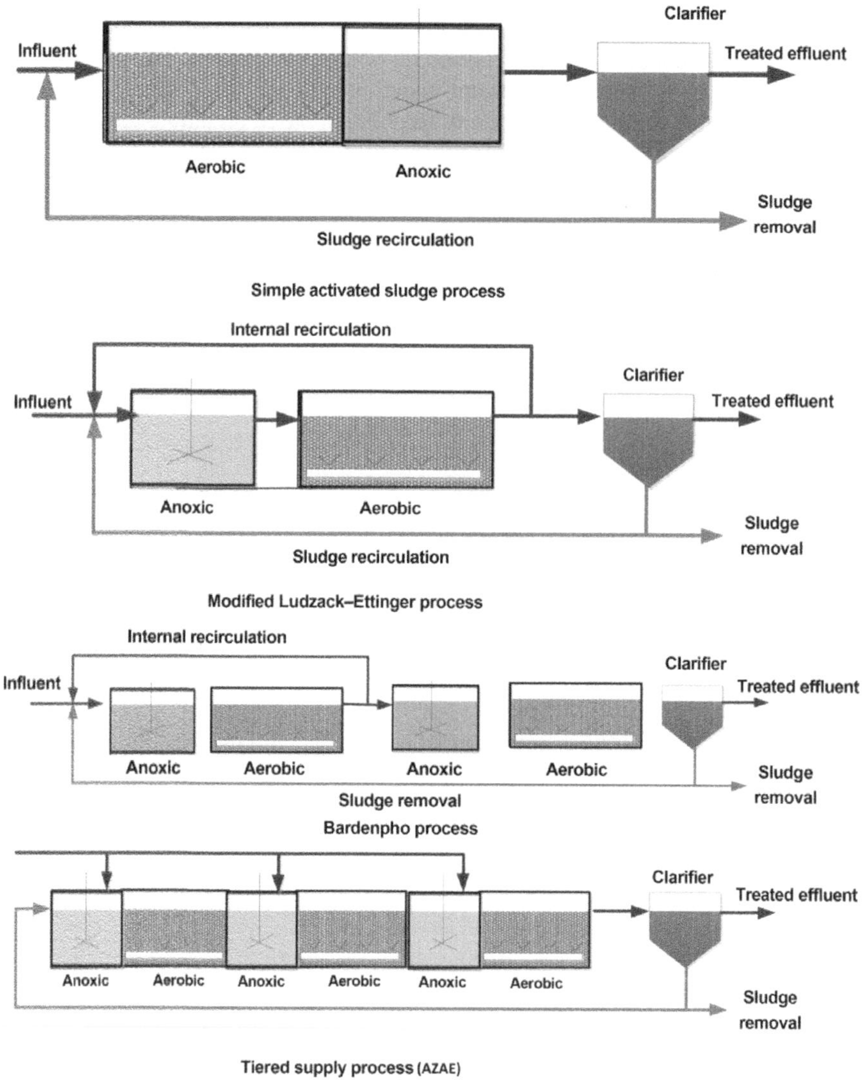

Figure 1.12A. *Some activated sludge nitrification–denitrification processes*

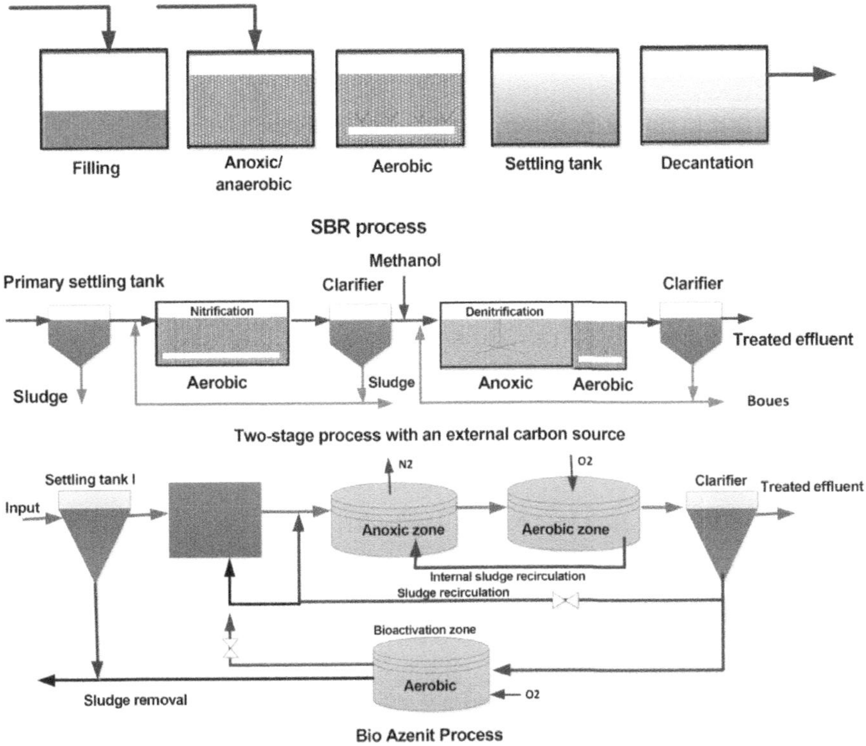

Figure 1.12B. *Some activated sludge nitrification–denitrification processes (continued)*

In practice, modernizing treatment plants involves adding a post-denitrification system followed by a final aeration stage to reoxygenate the effluent and refine the treatment process NH_4 (< 10 mg NGL.L^{-1}).

The main variants of the nitrification and denitrification processes combined in one or more reactors include:

– pre-denitrification (removal of nitrogen along with carbon from untreated wastewater);

– post-denitrification (elimination of nitrogen with carbon from endogenous respiration);

– alternating modes of aeration and anoxia in several stages;

– oxidation channels;

– batch sequencing reactors.

There are numerous denitrification processes. Figure 1.12 shows some of the processes proposed by manufacturers.

1.7. Autotrophic denitrification of activated sludge

The size of sulfur particles is an important parameter for sulfur–bacteria transfer. The smaller the size, the better the autotrophic denitrification process because the transfers are faster. The typical stoichiometric reaction of autotrophic denitrification with elemental sulfur involving *Thiobacillus denitrificans* is given in the following equation:

$$NO_3^- + 1.1\ S° + 0.4\ CO_2 + 0.76\ H_2 + 0.08\ NH_4^+ \rightarrow 0.5\ N_2 + 1.1\ SO_4^{2-} + 1.28\ H^+ + 0.08\ C_5H_7O_2N$$

The process of autotrophic denitrification with elemental sulfur can be presented as two consecutive reactions:

$$5S° + 6\ NO_3^- + 6H_2O \rightarrow 3\ N_2 + 5\ SO_4^{2-} + 4H_3O^+$$

$$11\ S° + 10\ NO_3^- + 4.1\ HCO_3^- + 0.5\ CO_2 + 1.72\ NH_4^+ + 2.54\ H_2 \rightarrow 0.92\ C_5H_7O_2N + 11\ SO_4^{2-} + 5.4\ N_2 + 9.62\ H^+$$

It can be noted that 2.51 g S° are consumed by g $N\text{-}NO_3^-$. Also, 7.54 $mg.L^{-1}$ sulfates are produced per 1 mg $N\text{-}NO_3.L^{-1}$ consumed (Gaid 1982). Different ratios (mg of sulfate produced per mg of denitrified $N\text{-}NO_3^-$) have been identified in the literature; namely, 1.5 (Liu 2009), 3.4 (Zhang 1999), 9.9 (Hashimoto 1987) and 11.1 (Schippers 1987).

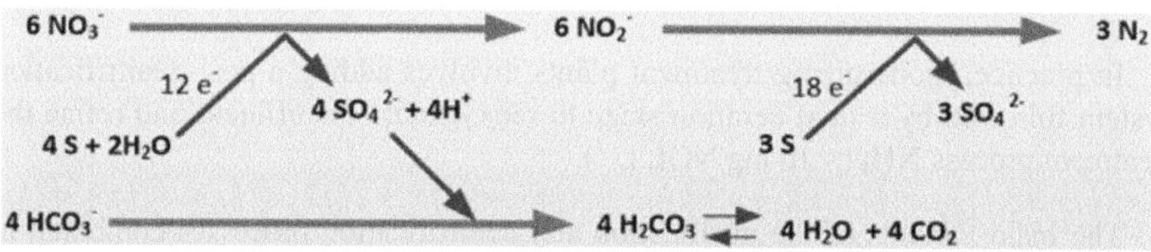

Figure 1.13. *Nitrate s-sulfur interactions and derivatives (Koenig and Liu 2001)*

Figure 1.13 summarizes the various chemical interactions occurring during autotrophic denitrification based on sulfur and limestone. According to this figure, *Thiobacillus denitrificans* acquire energy by oxidizing sulfur. Once the sulfur is oxidized, the released electrons will be used to reduce nitrates to nitrites, then to nitrogen gas.

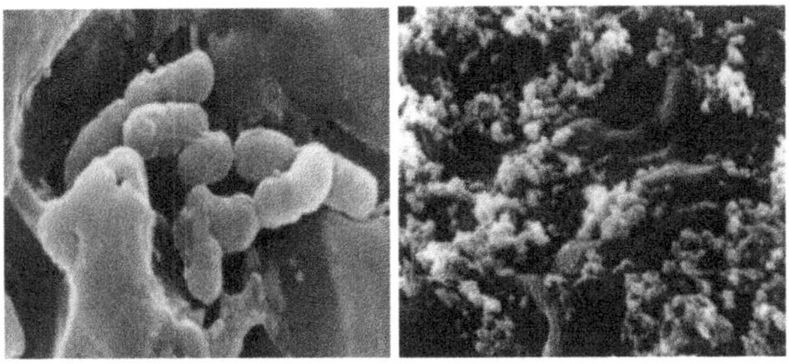

(Blais 1994) (Moon 2004)

Figure 1.14. *Sulfur particles colonized by bacterial strains containing Thiobacillus denitrificans*

1.7.1. *Autotrophic denitrification mechanism with sulfur*

Denitrification using sulfur compounds as electron donors is an alternative to heterotrophic denitrification for wastewater with high nitrate concentrations and low organic matter content. This process is carried out by autotrophic denitrifiers such as *Thiobacillus denitrificans* and *Thiomicrospira denitrificans*. The energy required by these microorganisms comes from redox reactions with elements such as hydrogen or various reduced sulfur compounds acting as electron donors. Autotrophic denitrifiers use inorganic carbon compounds, for example, as a carbon source for growth. Consequently, compared to conventional heterotrophic denitrification, autotrophic denitrification has two clear advantages:

– no need for an external organic carbon source, which reduces the cost of the process;

– lower biomass production, which minimizes sludge handling (Claus and Kutzner 1985; Zhang and Lampe 1999).

Current values for maximum growth rates of autotrophic denitrifying microorganisms (0.11 to 0.20 h^{-1} (Claus and Kutzner 1985; Oh 2000)) are similar to heterotrophic values (0.062 to 0.108 h^{-1} (Wiesmann 1994)), whereas biomass yields are lower for autotrophic denitrifying microorganisms (0.40 to 0.57 g of $VSS.g^{-1}$ (Claus and Kutzner 1985; Oh 2000)) compared to denitrifying heterotrophs (0.8 to 1.2 g of $VSS.g^{-1}$ (Wiesmann 1994)).

Under ambient conditions, S° exists as a ring formed of eight sulfur atoms (S8), forming a crystalline solid. Many S8 allotropes exist, but the most common is α-S8, which is essentially insoluble in water. Given its low solubility, it is unlikely that the microbial growth rates observed with S° can be obtained. It is thought that certain oxidizing bacteria produce extracellular enzymes that convert α-S8 into soluble polysulfides, Sx^{2-}, which can diffuse into a biofilm. Nanocrystalline S8 could also be an intermediate formed from Sx^{2-}. Other S°-oxidizing bacteria are known to have membrane-bound enzymes that obtain electrons by direct contact with the surface of S°, incorporating them directly into the cell.

The electron transfer mechanism is important because it can impact biofilm development and its kinetic behavior. If electron transfer is limited to bacteria in direct contact with S°, increasing biofilm thickness will not increase denitrification fluxes (the rate at which So removes NO_3^- per unit area of S°).

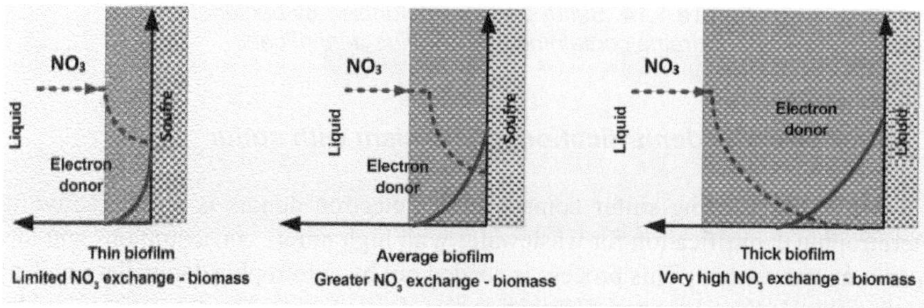

Figure 1.15. *Sulfur denitrification depending on biofilm thickness*

However, if soluble sulfur species diffuse into the biofilm, the increased biofilm thickness would provide more active biomass. This would allow for larger NO_3^- fluxes. The biofilm developing on S° is subject to two opposing diffusions: the donor and acceptor diffuse into the biofilms from opposite zones (see Figure 1.15). The kinetic behavior of counter-diffusion biofilms differs from that of conventional co-diffusion biofilms. Nitrate fluxes increase as biofilm thickness increases but eventually decrease when the film becomes too thick.

The decline is due to limited diffusion, because the donor and acceptor enter the biofilm from opposite sides and therefore have longer diffusion distances and thus lower concentrations when they meet (Essila 2000; Martin and Nerenberg 2012).

However, the form of sulfur (e.g. α-S8, polysulfides, S8 nanocrystalline) and the identity of the true electron donor are not fully explained. It is also unclear whether the soluble sulfur is only available to bacteria on the surface of the solid or whether it diffuses into the biofilm. This uncertainty and the complexity of counter-diffusion explain why the denitrification kinetics based on S° remain poorly characterized compared to other electron donors.

Given that the mechanisms of electron transfer from S° are not well established, the best approach might be to characterize denitrification rates as a function of biofilm development and NO_3^- concentration.

As a biofilm process, the S°-based denitrification rate should be related to the surface area of the S° rather than the S° concentration (Sierra-Alvarez et al. 2007). This rate should vary depending on both the NO_3^- concentration and biofilm thickness.

1.7.2. *Autotrophic denitrification in free culture (activated sludge)*

Factors affecting the performance of autotrophic denitrification are those that directly impact microbial growth and activity, such as pH, temperature, dissolved oxygen, the presence of organic or inorganic contaminants and byproducts (nitrites, sulfates). Other parameters, such as sulfur and limestone particle size, the sulfur/limestone ratio, the nitrate concentration of the influencing material, the treatment flow rate and the hydraulic retention time, are also interrelated.

1.7.2.1. *Temperature*

The optimal temperature for the growth of denitrifying bacteria is approximately 30°C. Amatya et al. (2005) studied the effect of temperature on biological denitrification and established that the denitrification rate is significant above 16°C and maximal at 30°C, with a nitrate removal efficiency of 76.9% in columnar reactors. Temperature not only plays an important role in bacterial cell growth but also has a direct effect on gene expression of enzymes involved in denitrification. Maximum denitrification was reached at temperatures ranging between 33 and 35°C (Oh et al. 2000). However, another study showed that it was possible to completely denitrify all nitrates at 10°C, except that the duration of the treatment is significantly longer compared with the other temperatures mentioned previously. Trouve et al. (1998) isolated cryophilic strains of *Thiobacillus denitrificans* capable of denitrifying at temperatures as low as 5°C with denitrification efficiency similar to that occurring at 20°C. The problem with low temperatures lies in the fact that the denitrification process would require longer incubation periods. Complete nitrate removal at 10°C requires 36 to 78 hours, and the same strains may require approximately 48 to over 120 hours of incubation using a batch process.

Furthermore, depending on the sulfur source used, it appears that at a temperature of 5°C and a pH between 6 and 8, nitrate removal is more efficient in the presence of thiosulfate compared to elemental sulfur. A classification describing the affinity of *Thiobacillus denitrificans* regarding the available sulfur source has been proposed: $S_2O_3^{2-} > FeS > FeS_2 > S°$ (Ben Khaled 2016).

1.7.2.2. *pH*

Autotrophic denitrification with sulfur leads to acidification of the environment, caused by a decrease in pH and alkalinity. The reduction of one mole of nitrate produces 1.10 moles of H^+ and leads to bicarbonate consumption. The pH parameter is important for the growth and survival of any microorganism, and since autotrophic denitrification is regulated by microorganisms, pH plays a critical role in the NO_3-S° system. The optimal pH for most autotrophic denitrifying bacteria ranges between 6 and 9 (Holt et al. 1994). However, a study by Koenig also demonstrated that autotrophic denitrifying bacteria such as *Thiobacillus denitrificans* have an optimal pH in the range of 6.8 to 8.2 and that a pH below 6 is detrimental to the process (Koenig and Liu 2002). At low pH (below 6), nitrogen oxide reductases are inhibited, particularly those reducing nitrous oxide (N_2O) and the overall denitrification rate decreases due to the accumulation of N_2O (Bergaust et al. 2010). In order to achieve complete denitrification, the pH of the system must be neutralized with suitable substrates such as sodium bicarbonate.

1.7.2.3. *Dissolved oxygen*

Dissolved oxygen (DO) poses a major problem in denitrification mechanisms, whether with heterotrophic or autotrophic bacteria. Since it is part of the basic nutritional requirements of microorganisms, oxygen is preferentially assimilated compared to nitrates present in the effluents to be treated. Furthermore, the electrochemical potential of oxygen is higher than that of nitrates. Consequently, oxygen competes with nitrates as the final electron acceptor and inhibits denitrification.

1.7.2.4. *Influence of the S/N ratio*

The effect of different S/N ratios on the efficiency of autotrophic sulfur denitrification was studied by Campos et al. (2008). They varied the concentrations of thiosulfate as a sulfur source relative to that of nitrates. When the S/N ratio increased from 1.16 to 2.44 g $S\text{-}S_2O_3.g^{-1}N\text{-}NO_3$, an accumulation of nitrites was observed, and denitrification could not be completed. However, with a higher S/N ratio (6.67 $g.g^{-1}$), nitrates were completely reduced to nitrogen gas at the start of the experiments, but nitrites still accumulated afterwards (Ben Khaled 2016).

1.7.2.5. *Influence of the S°/X ratio*

The specific rate of denitrification is given by:

$$\text{vDN} = \frac{1}{\text{X}}\frac{\partial N}{\partial t}$$

where X is the concentration of biomass in the reactor (mg.L^{-1}). This relationship shows that v_{DN} is proportional to the S°/X ratio when this ratio is less than 50 g.g^{-1}. Under constant operating conditions (pH between 7.5 and 8; alkalinity between 270 and 280 mg $CaCO_3$; temperature = 20°C), the denitrification kinetics show the evolution of the various elements (nitrates, pH, sulfates), as represented by a few examples (see Figure 1.16). These results allow us to calculate the denitrification rates v_{DN} (mg N-NO_3^-.g.MVS^{-1}.h^{-1}) depending on the S°/X ratio.

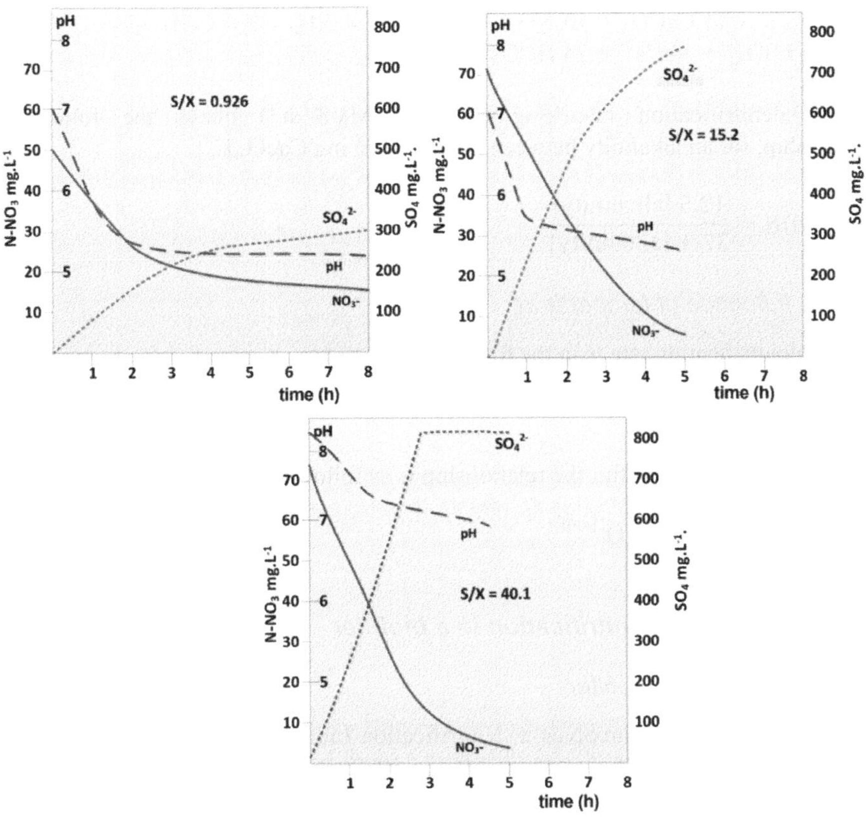

Figure 1.16. *Influence of S°/X ratio on sulfur denitrification*

We obtain the following:

$$\text{vDN} = \frac{5\ \text{S°/X}}{9.1 + \text{S°/X}}$$

The rate of nitrate–sulfur denitrification follows a Monod-type law. It depends on the ratio of sulfur to sludge present at the initial moment of the reaction.

1.7.2.6. *Influence of alkalinity and alkalinity ratio*

According to the sulfur denitrification reaction, autotrophic denitrifying bacteria require 3.91 g of alkalinity in the form of $CaCO_3$ per gram of reduced N-NO_3 to control the pH:

$$55\ S + 44\ CaCO_3 + 50\ NO_3^- + 18\ H_2O + 4\ NH_4^+ \rightarrow 4\ C_5H_7O_2N + 25\ N_2 + 55\ SO_4^{2-} + 44\ Ca^{2+} + 24\ HCO_3^-$$

The denitrification rate (mg N-NO_3.g $MVS^{-1}.h^{-1}$) obeys the following relationship, for an alkalinity between 200 and 340 mg $CacO_3.L^{-1}$:

$$vDN = \frac{12.5\ [alkalinity]}{37 + [alkalinity]}$$

1.7.2.7. *Influence of temperature*

The denitrification rate is in the form:

$$v_{DN} = ks\ \Theta^{(T-10)}$$

In the presence of sulfur, the relationship is as follows:

$$v_{DN} = 2.01\ x\ 1.12\ \Theta^{(T-10)}$$

1.7.3. *Autotrophic denitrification in a biofilter*

1.7.3.1. *Submerged biofilter*

The Veolia process involves a denitrification bioreactor filled with a granular mixture of elemental sulfur and maerl (Gaid 1982; Ratel and Debrieu 2003). Upon exiting the columns, the water passes through post-treatment systems for filtration, reoxygenation and disinfection. This is done to avoid excessive prolonging of the seeding period of the columns with the autotrophic denitrifying bacteria *Thiobacillus denitrificans*. Due to the low growth rate, the system is enriched with phosphates, nitrates, ammonia and bicarbonates by recirculating the enrichment solution for two weeks. To prevent pressure losses due to increased biomass, backwashing with air is performed every 10 days. The process has some drawbacks related to the significant amount of sulfates produced and the need for low flow rates to optimize denitrification efficiency.

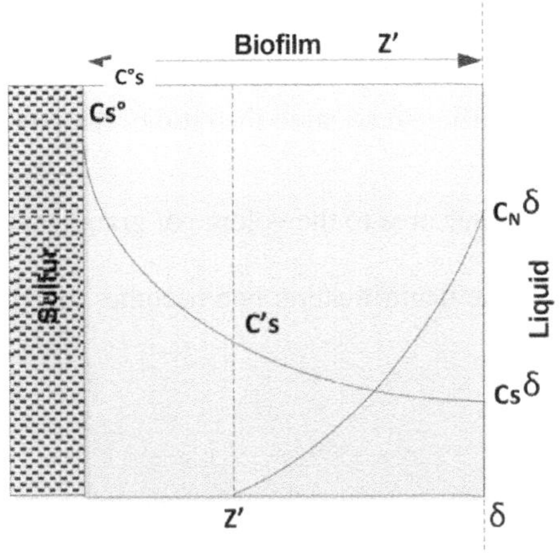

Figure 1.17. *Evolution of the various elements in a sulfur–nitrates–biomass system*

Figure 1.17 shows the evolution of sulfur and nitrates in a biofilm developed with sulfur. Cs and C_N represent the concentrations of S and N-NO_3 in the biofilm, respectively. Z is the distance in the film from the surface S. δ is the thickness of the biofilm. Sulfur (C°s) dissolves in the biofilm at a constant rate. This concentration decreases by C°s at Cs during passage through the biofilm. Due to its insolubility in water, sulfur does not penetrate the water. Nitrates C_N^{δ} are reduced by passage through the biofilm to a zero concentration at the point Z = Z′. Assuming homogeneity of the biofilm, establishing a mass balance leads to the following denitrification equation. For a low ratio S/X:

$$\mathrm{vDN} = \frac{\mathrm{ks}}{\mathrm{Y}} \mathrm{C°s} \left(\frac{\mathrm{Ds}}{\mathrm{ks}}\right)^{1/2} \frac{\rho \mathrm{b}}{\rho \mathrm{S}} \left(\frac{\mathrm{A}}{\mathrm{V}}\right) \frac{\mathrm{Cs}}{\mathrm{X}}$$

with:

– ks: first-order rate constant in intrafilm reactions (d^{-1});

– C°s: sulfur concentration at the start of the biofilm ($M.L^{-3}$);

– Ds: sulfur diffusion coefficient ($L^2.t^{-1}$);

– ρb: sludge density ($M.L^{-3}$);

– ρs: sulfur density ($M.L^{-3}$);

– A: biofilm area (L^2);

– Cs: sulfur concentration in the reactor ($M.L^{-3}$);

– X: biomass concentration in the reactor ($M.L^{-3}$);

– Y: stoichiometric coefficient equal to the ratio of the rate of sulfur consumption to the denitrification rate;

– A/V: ratio of the surface area to the volume of granular sulfur ($L^2.L^{-3}$).

For a high S/X ratio, the denitrification rate becomes independent of sulfur and is written as:

$$vDN = \frac{ks}{Y} C°s \frac{\rho b}{\rho S}$$

The Monod saturation constant (Ks) is low (Claus and Kutzner 1985; Batchelor 1987). The reaction that takes place within the biofilm is zero order. Because the substrate penetrates the biofilm pores slowly, the zero-order reaction within the biofilm becomes a half-order reaction at the biofilm surface (Koenig and Liu 2001).

The flow passing through the biofilter is considered a piston flow. We can write:

$$\frac{\partial C}{\partial \delta} = -v'DN \frac{A}{Q}$$

where C is the concentration, δ is the thickness of the biofilm, Q is the flow rate, A is the surface area traversed by the flow and v'DN is the denitrification rate per unit volume of the biofilter.

Since the reaction in the biofilm is half-order:

$$v'DN = k\, C^{1/2}$$

Replacing v'DN, it becomes:

$$-\frac{\partial c}{\partial t} = -k\sqrt{C}$$

where k is the half-order reaction constant per unit volume of biofilter; i.e.:

$$\sqrt{Ct} = \sqrt{Co} - \frac{kt}{2}$$

1.7.3.2. *Influence of volumetric load*

The influence of volumetric load (kg $N\text{-}NO_3.m^{-3}$ sulfur. d^{-1}) on the denitrification yield (R) is given by the relation:

$$R\,(\%) = a\,Cv^{b}$$

where a and b are constants related to R and the percentage of sulfur contained in the biofilter.

Table 1.9 summarizes the values of a and b.

Sulfur (%)	R (%)
5	$R = 0.29\ Cv^{-1.45}$
15	$R = 0.70\ Cv^{-1.40}$
30	$R = 0.90\ Cv^{-1.13}$
45	$R = 1.05\ Cv^{-0.82}$
100	$R = 1.23\ Cv^{-0.68}$

Table 1.9. *Denitrification efficiency as a function of sulfur load*

1.7.3.3. *Influence of alkalinity*

The influence of alkalinity on the denitrification rate (mg $N\text{-}NO_3$ removed.h^{-1}) is indicated by the following:

$$\text{vDN} = 125\ \frac{\text{alkalinity}}{41.6 + \text{alkalinity}}$$

Alkalinity is expressed in °f:

$$1°f = 10\ \text{mg}\ CaCO_3.L^{-1}$$

1.7.3.4. *Influence of the sulfur/limestone ratio*

Alkalinity is an important parameter that regulates denitrification kinetics. Carbonate consumption linked to the formation of H^+ ions tends to rapidly lower the pH and limit bacterial activity.

Because the biological reaction of autotrophic denitrification with sulfur generates H^+ ions, the process consumes alkalinity and therefore requires a buffer substrate to prevent the pH from dropping. Limestone is used as the solid phase to neutralize the acidity generated during the autotrophic denitrification reaction with sulfur (Batchelor 1978; Gaid 1982). Batchelor (1978) and Gaid (1982) showed that limestone is a well-suited neutralizing agent for autotrophic denitrification with sulfur and that the ratio of initial alkalinity to the theoretically required alkalinity in wastewater must be at least 0.5 for effective nitrate removal. The sulfur-to-limestone ratio is generally determined by the alkalinity of the tributary. Alkalinity is the ability of water to maintain a stable pH (buffering capacity) and is reflected by the presence of hydroxide ions (OH^-), bicarbonate ions (HCO_3^-) and carbonate ions (CO_3^{2-}). Therefore, consuming bicarbonate leads to a decrease in alkalinity.

Theoretically, during an autotrophic denitrification process, 1 mg of denitrified N-NO_3 leads to the consumption of 4.5 mg of $CaCO_3$ (Sengupta et al. 2007). However, other authors have reported relatively different results, indicating that 1 mg of denitrified N-NO_3 would consume 5.6 mg of $CaCO_3$ (Zhang and Lampe 1999) and 4.77 mg of $CaCO_3$ (Moon 2006). During autotrophic denitrification, the dissolution of limestone to neutralize acidity results in the transformation of bicarbonate into carbonic acid in the presence of H^+ ions, thus generating the release of carbon dioxide.

The neutralization step proceeds according to the following equations:

$$CaCO_3(s) + H^+ \rightarrow HCO_3^- + Ca^{2+}$$

$$HCO_3^- + H^+ \rightarrow H_2CO_3$$

$$H_2CO_3 \rightarrow H_2O + CO_2$$

A 1:1 S:C ratio ensures a nitrate removal efficiency of approximately 98% (Gaid 1982). A 1:1 ratio would be optimal to achieve a high nitrate removal efficiency with low sulfate production (Sikora 1976; Schippers 1987). Certain factors such as affluent pH, the initial alkalinity and the limestone requirement can influence the optimal S:C ratio via denitrification (Liu 2002). Indeed, if the initial alkalinity is present in the form of bicarbonate and the pH of the affluent to be treated is around 8, an S:C ratio greater than 2:1 is sufficient to ensure effective denitrification. This requirement depends largely on the minimum pH allowed for maintaining denitrification, which determines the S:C ratio. Therefore, the optimal S:C ratio would be defined as the proportion at which the volumetric loading is optimal.

1.7.3.4.1. Sulfur and limestone configuration in the bioreactor

The lining configuration of the denitrifying biofilter varies depending on the process applied and the type of wastewater to be treated. The biofilter lining can have several configurations in which granular elemental sulfur and granular limestone are mixed. One original idea is to use mixtures of sulfur with calcium carbonate. This solution is simple to implement with marine maerl, which consists of 80 to 85% calcium carbonate, 10 to 15% magnesium carbonate, 0.65% calcium phosphate, 0.60% sulfur and 1% organic nitrogen (Gaid 1982).

The applied volumetric loading rates are higher due to the continuous pH neutralization by the maerl (see Table 1.10 and Figure 1.18). The total filtration height is 1.75 m.

	Composition of the biofilter		
	Sulfur 50% Maerl 50%	**Sulfur 45% Gravel 55%**	**Sulfur 100%**
Material dimensions (mm)	Sulfur: 2–5 Maerl: 2–5	Sulfur: 2–5 Gravel: 3–7	Sulfur: 2–5
Volumetric load ($kg\ N\text{-}NO_3$ applied.m^3. d^1)	3	1	1.85

Table 1.10. *Biofilter performance with different material compositions*

The majority of studies cited in the literature refer to a DASC where sulfur and limestone are mixed in the same bioreactor, generally referred to as a "fixed-bed reactor" (Gaid 1982; Zhang 2004; Sengupta et al. 2007; Zhou et al. 2011).

The autotrophic denitrification kinetics with sulfur follow a half-order reaction and imply that the denitrification efficiency is dependent on the specific surface area of the reactive sulfur particle (Koenig and Liu 2001).

The autotrophic denitrification process can use electron donors such as S°, sulfide (S^{2-}), SCN^-, thiosulfate ($S_2O_3^{2-}$), sulfite (SO_3^{2-}), metallic elements and ions. The biomass yield of autotrophic denitrifiers is generally lower than that of heterotrophic denitrifiers, which is advantageous due to reduced sludge management.

However, metallic elements and ions, as well as sulfur and its compounds, generate harmful chemicals (Di Capua et al. 2019). S^{2-} can exist in water in the form of molecular hydrogen sulfide (H_2S) or hydrogen sulfide ions (HS^-)/sulfide (S^{2-}) according to the medium conditions (Wiemann et al. 1998; Di Capua et al. 2019). S^{2-} presents serious environmental challenges due to its toxicity, corrosivity and odor (Di Capua et al. 2019). Furthermore, it is an inhibitor for many microorganisms when present in high concentrations, including autotrophic denitrifying bacteria (Cardoso et al. 2006).

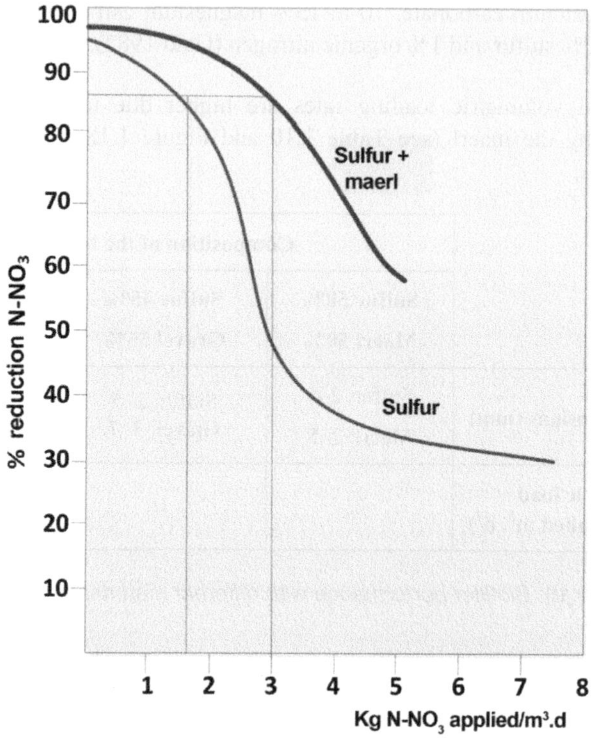

Figure 1.18. *Denitrification efficiency as a function of the applied load for two sulfur loads*

The two-step process is the preferred method for autotrophic denitrifiers using S, probably due to lower electronic requirements compared to direct oxidation of S^{2-} into SO_4^{2-} (Moraes et al. 2012). The formation of S as an intermediate product could lead to the formation of a white/yellowish color that would disappear when the S is subsequently consumed. Furthermore, the consumption of S leads to a fluctuation in alkalinity since the oxidation of S into SO_4^{2-} leads to a decrease in carbonates and a drop in pH.

1.8. Nitrogen treatment by anammox bacteria

The oldest and still most widespread biological process for removing nitrogenous species from wastewater is nitrification–denitrification, that is, the nitrification of ammonia into nitrate, followed by an organic reduction of nitrate into diatomic nitrogen. For sustainability reasons, particularly due to the energy required to supply oxygen and the amount of organic matter needed, which could otherwise be used to produce methane fuel, other potential nitrogen removal processes have emerged. Among these, biological processes have been implemented, using a unique group of bacteria called "anammox," which anaerobically convert ammonia into nitrogen gas while using nitrite as an electron acceptor. Thus, the ammonia is oxidized solely to nitrite and then denitrified. This reduces both the oxygen and organic matter requirements for denitrification.

In this process, only about half of the ammonia is initially oxidized to nitrite by conventional ammonia-oxidizing bacteria. Anammox bacteria then use this nitrite as an electron acceptor under anoxic conditions to oxidize the remaining half of the ammonia, converting both into nitrogen gas.

Classically, ammonia is oxidized to nitrite by ammonia-oxidizing bacteria (AOB), and the biochemical process is carried out by the *Nitrosomonas* and *Nitrospira*. Next, nitrites are oxidized to nitrates by nitrite-oxidizing bacteria (NOB) with the help of microorganisms such as *Nitrobacter* and *Nitrospira*. The denitrification process is primarily carried out by heterotrophic bacteria, such as *Paracoccus denitrificans* and *Pseudomonas stutzeri*, which require organic carbon resources as electron donors to reduce nitrogen oxides to nitrogen gas.

Due to the oxygen supply required for the nitrification process and the organic matter requirements for the denitrification process, the conventional activated sludge process is an energy-intensive process.

New bacterial communities have been discovered, which show advantages over conventional nitrifying or denitrifying bacteria, such as higher affinity for the substrate, better physicochemical tolerances and/or lower greenhouse gas emissions. Overall, these new microbial communities could offer new possibilities for efficient and energy-saving nitrogen removal from wastewater. These microbial communities are capable of using other electron acceptors or donors, and of offering substantial advantages over the classical nitrogen removal process, such as lower production of N_2O, energy savings during aeration and organic carbon savings during anoxic denitrification. In addition, they can better cope with adverse environments such as low O_2 or ammonia deficiency conditions.

The anammox (anaerobic ammonium oxidation) process is a new pathway in the nitrogen cycle (see Figure 1.19). This process involves oxidizing ammonium under anaerobic conditions using nitrite as an electron acceptor instead of oxygen. It results in the final production of N_2 (inert gaseous product) and nitrate. It is carried out by autotrophic microorganisms not yet cultured in pure strains but for which enrichment conditions are known in laboratory reactors. The anaerobic oxidation of ammonium (anammox) is a new metabolic pathway (Lam et al. 2009) in the nitrogen cycle, offering new opportunities for process engineers and microbiologists. The existence of these bacteria capable of carrying out the anaerobic oxidation of ammonium (anammox), with nitrates and/or nitrites, to nitrogen gas opens up new perspectives because they bring simplifications to the treatment processes.

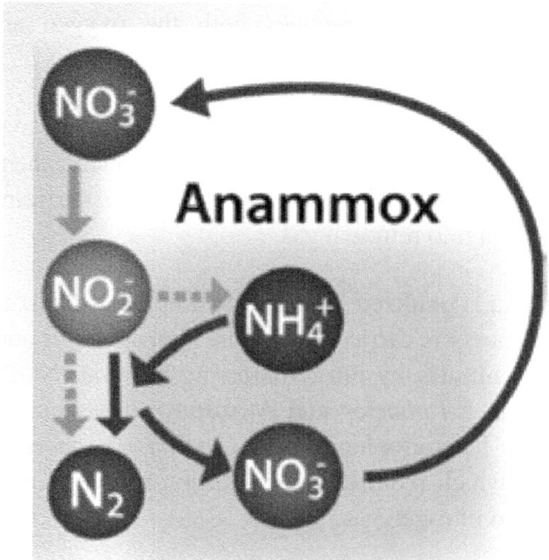

Figure 1.19. *Simplified diagram of the desammonification and denitrification process*

1.8.1. *Anammox characteristics*

The anammox process is currently linked to a monophyletic group of bacteria belonging to the phylum of *Planctomycetes*. The bacteria that carry out this process are still poorly understood, but are quite unique.

1.8.2. *Microbiological characteristics*

Anammox bacteria are cells approximately 1 µm in diameter with a cell envelope organization very different from that of Gram-negative or Gram-positive bacteria. They consist of a cell wall lacking peptidoglycan, a cytoplasmic membrane and an intracytoplasmic membrane. Their cytoplasm is divided into three compartments: the paryphoplasm, the riboplasm and the anammoxosome (Niftrik et al. 2004). This last compartment is specific to anammox bacteria.

They contain a specific intracellular compartment, the anammoxosome, which occupies a significant volume of the cell and where the anammox process takes place. The membrane of this vesicle is made up of lipids with a particular chemical structure, ladderanes, consisting of three or five fused cyclobutane rings. These molecules with constrained geometry form an extremely dense barrier, thus greatly reducing the membrane's permeability to small molecules, particularly hydrazine (N_2H_4), an intermediate and highly toxic product of the anammox reaction (Kuenen 2008).

Currently, five genera of anammox bacteria have been (provisionally) defined: *Brocadia*, *Kuenenia*, *Anammoxoglobus* and *Jettenia* regarding freshwater species, as well as Scalindua for marine species.

In summary, anammox bacteria are characterized by several remarkable properties:

– They all possess an anammoxosome, a large membrane-bound compartment within the cytoplasm where anammox catabolism takes place.

– The lipids constituting the membranes of these bacteria contain a high proportion of ladderane-type fatty acids, such as pentacycloanammoxic acid, which are currently unique in biology.

– Hydrazine N_2H_4 and hydroxylamine NH_2OH, toxic to most living beings, are metabolic intermediates in the anammox process.

– Their growth rate is extremely slow, with a doubling rate of nearly two weeks. These microorganisms are strict anaerobes and chemolithoautotrophs (Niftrik 2004). Their doubling time is approximately 1.8 to 11 days (Magrí 2012), their cell yield is very low (0.11 g $VSS.g^{-1}$ $N\text{-}NH_4^+$) and they are very sensitive to variations in environmental conditions, which makes them extremely difficult to cultivate (Strous et al. 1999; Ni and Meng 2011).

Methods for obtaining pure cultures of Anammox bacteria are complex and require very strict protocols. However, enrichment cultures are available, and everything we know about them derives from these enrichments (Dalsgaard et al. 2005).

1.8.3. *Biochemistry of anaerobic oxidation of ammonium*

The biochemistry of anammox bacteria is not yet fully understood. It is accepted that the anammox process takes place in the anammoxosome and that its objectives are the synthesis of adenosine triphosphate (ATP) and carbon fixation (Niftrik et al. 2004). According to Figure 1.20, the proposed mechanism indicates that nitrites are reduced to hydroxylamine (NH_2OH) by a nitrite reductase (NIR). Ammonia and hydroxylamine are then used by hydrazine hydrolase to producc hydrazine (N_2H_4).

Hydrazine oxidoreductase oxidizes hydrazine into N_2 and releases four protons and four electrons. The four released electrons are used with five protons from the riboplasm by nitrite reductase (Nir) to complete the catalytic cycle. All of these reactions establish a proton gradient, which is used for ATP synthesis catalyzed by adenosine triphosphatase (ATPases) bound to the anammoxosome membrane. The synthesized ATP is then released into the riboplasm (Jetten et al. 1999; Schalk 2000; Jetten et al. 2001; Brandes et al. 2007).

The reactions are as follows:

$$NO_2^- + 5H^+ + 4es \rightarrow NH_2OH + H_2O$$

$$NH_2OH + NH_3 \rightarrow N_2H_4 + H_2O$$

$$N_2H_4 \rightarrow N_2 + 4H^+ + 4es$$

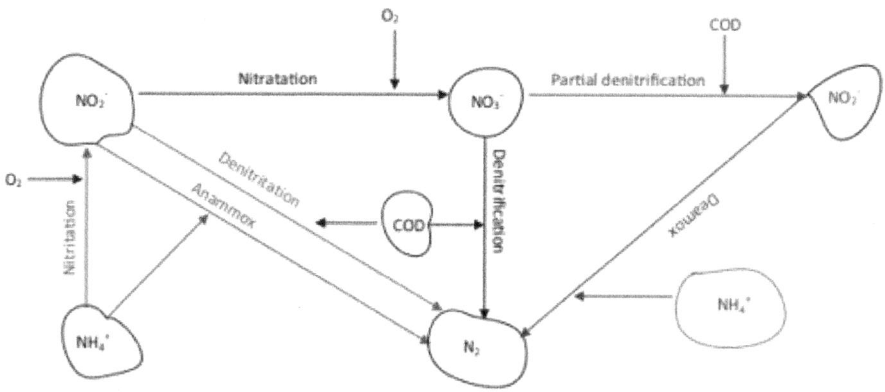

Figure 1.20. *Overall diagram of anammox involvement (Kosgey et al. 2022)*

Excess nitrite (0.3 mol/mol ammonium) is oxidized anaerobically to nitrates. The electrons from this oxidation are probably used for CO_2 fixation (Van de Graaf et al. 1995).

The overall assessment of reactions, proposed by Strous et al. (1999), is presented here:

$$NH_4^+ + NO_2^- = N_2 + 2H_2O \text{ (x 15)}$$

$$CO_2 + 2\ NO_2^- + H_2O = CH_2OH + 2\ NO_3^-$$

$NH_4^+ + 1.32\ NO_2^- + 0066HCO_3^- \rightarrow 1.02N_2 + 0.256\ NO_3^- + 0.066CH_2O_{0.5}N_{0.15} + 2.03H_2O$

The catabolic reaction is replicated 15 times to bind a molecule of carbon dioxide with nitrites. These, by releasing electrons, lead to the anaerobic production of nitrate.

1.8.4. *Factors influencing the growth of anammox bacteria*

To date, five genera of anammox bacteria have been identified in the literature, forming the monophyletic group Brocadiales (Ali et al. 2013). These different genera appear to be able to occupy different ecological niches. The metabolism of anammox bacteria allows them to colonize habitats and adapt to new situations by using a variety of electron donors and acceptors (e.g. manganese, iron oxides and formate) (Strous et al. 2006). Indeed, these bacteria regularly compete with other bacteria in the nitrogen cycle (nitrifying and denitrifying bacteria) and with archaea due to biological interactions between these microbial groups for nitrogen and the CO_2.

A potential for specific niche differentiation has thus been observed for each genus of Anammox in natural and anthropogenic systems, and even dominance in certain environments (such as oxygen-deprived areas). The most obvious examples include marine environments, wastewater treatment plants, freshwater, estuaries and deep-sea hydrothermal vents (Hassan 2018).

These observations have led to numerous hypotheses regarding the environmental conditions conducive to the enrichment of anammox bacteria.

1.8.5. *Salinity*

Brummer suggests that the species of *Planctomycetes* in freshwater environments differ from those found in marine environments (Brummer et al. 2004). In a number of estuaries, the anammox level decreases with increasing salinity (Trimmer et al. 2003). The type of anammox communities is changing, favoring the growth of genera best adapted to increased salinity (Kartal et al. 2006; Hassan 2018).

1.8.6. *Presence of organic matter and/or nutrients*

A change in environmental conditions (e.g. for the start-up of a bioreactor for anammox enrichment) can also trigger a change in the dominant anammox genus (Van Der Star et al. 2007).

Nitrite and micronutrient concentrations, the use of alternative energy sources, the presence of carbon sources (propionate or acetate) (Kartal et al. 2007, 2008) as well as anthropogenic inputs of nutrients can lead to marked changes in the structure of the community (Li et al. 2011; Hassan 2018).

1.8.7. Candidatus Brocadia *and* Kuenenia

These species have attracted considerable attention because they are the genera commonly found in wastewater treatment plants. The niche difference between these two genera is still unknown. Thus, the change in dominance over time of *Brocadia* towards *Kuenenia* was observed in a membrane biofilm reactor (MBR) (Van Der star et al. 2008) and in batch reactors. In both cases, the affinity for nitrite was assumed to be the limiting factor.

Based on these observations, the authors hypothesized that anammox bacteria have very low growth rates, and therefore, long startup periods are necessary for their enrichment. Periods of 58 days (Liao et al. 2007) and 60 days (Dapena-Mora et al. 2007) have been reported in batch bioreactors (SBRs) using methanogenic granular sludge or activated sludge as the inoculum.

Table 1.11 summarizes the growth and functioning parameters of various anammox bacteria.

In large-scale installations, generation times of 11 days (Van Der Star et al. 2007) and 27 days (Joss et al. 2009) were measured. These characteristics make the implementation of anammox bacteria in a treatment process challenging in terms of achieving enrichment and then maintaining activity in the process.

Strategies have been reported in the literature aimed at reducing startup time and/or promoting biofilm formation (Fernández et al. 2008; Hassan 2018), including the use of membrane bioreactors (Van Der Star et al. 2008).

However, exposure to inhibitory conditions could lead to complete reactor failure, regardless of the strategy used to conserve biomass and maintain anammox bacteria. Therefore, restoring full wastewater treatment capacity would be a very lengthy process. Very little information is available on the sensitivity of Anammox bacteria to compounds commonly found in wastewater. Substrate concentration, oxygen levels, temperature, pH, etc., must be considered for the application and industrialization of the process (Hassan 2018).

Parameters	***Brocadia anammoxidans***	***Brocadia sinica***	***Kuenenia stuttgartiensis***	***Scalindua* sp.**	***Scalindua profunda***	***Jettenia caeni***
Growth temperature (°C)	20–43	25–45	25–37	10–30	15–45	20–42.5
Growth pH	6.7–8.3	7.0–8.8	6.5–9.0	6.0–8.5	7.4	6.5–8.5
Max growth rate (μmax, d^{-1})	0.065	0.098	0.062–0.084	0.048	–	0.048
Doubling rate (d)	10.7	7	8–11	–	–	–
Cell yield am (mmol C mmol N^{-1})	0.07	0.063	–	0.03	–	0.056
km per NH_4 (μM) (5*m°)	< 5	28 +/- 4	–	3	–	17.1 +/- 4.3
km per NO_2 (μM)	< 5	34 +/- 21	0.2–3	0.45	–	35.6 +/- 0.92

Table 1.11. *Physiological and kinetic characteristics for anammox bacteria (Ni et al. 2016)*

1.8.8. *Oxygen*

Therefore, restoring full wastewater treatment capacity would be a very lengthy process. Very little information is available on the sensitivity of anammox bacteria to compounds commonly found in wastewater. Substrate concentration, oxygen levels, temperature, pH, etc., must be considered for the application and industrialization of the process (Jung et al. 2007). It has been reported that the anaerobic oxidation process of ammonium is reversibly inhibited by dissolved oxygen at a low level (< 1% of air saturation) (Strous et al. 1997). Conversely, irreversible inhibition occurs at high concentrations (> 18% of air saturation) (Egli et al. 2001). Therefore, dissolved oxygen must be strictly controlled in processes to prevent process shutdown and protect the maintenance of anammox bacteria.

1.8.9. *Temperature*

The maximum activity of anammox was observed between 35 and 40°C in a sequential batch reactor (Dosta et al. 2008). Egli et al. (2001) reported an optimal temperature of 37°C for anammox. Temperatures above this range cause an irreversible drop in anammox activity due to cell lysis, while temperatures below 15°C make the process unstable (Dosta et al. 2008).

In marine environments where *Candidatus scalindua* is the only type detected, optimal anammox activity was measured at 12°C (Rysgaard et al. 2004). These results suggest that anammox activity differs depending on the genus considered, as well as that an anammox process could be acclimatized and then used to treat wastewater at different temperatures.

1.8.10. *Substrate concentrations*

Anammox bacteria can be inhibited by their own substrates, nitrites and ammonium ions. Generally, they are more sensitive to high nitrite concentrations than to high ammonium ion concentrations. It is important to control the concentrations of these substrates in bioreactors and to adjust their stoichiometry in the flux, even though the literature does not offer a consensus regarding the limiting values (thresholds) of inhibition.

1.8.11. *Nitrite*

Numerous studies have confirmed that nitrite is a feedback inhibitor of anaerobic ammonium oxidation when it reaches excessively high concentrations. However, the observed inhibition thresholds vary between 48 and 750 mg N-NO_2.L^{-1} according to the operating procedures.

These threshold differences can be attributed to differences in biomass characteristics (anammox species) and/or operational conditions (pH, reactor configuration, temperature, substrate concentrations, etc.).

Anammox bacteria can be inhibited by their own substrates, nitrite and ammonium. Generally, they are more sensitive to high nitrite concentrations than to high ammonium concentrations. It is important to control the concentrations of these substrates in the bioreactors and to adjust their stoichiometry in the influencing system.

In an SBR containing aggregated anammox biomass (80% anammox bacteria/total bacteria), an immediate and complete inhibition of the anammox process was observed at nitrite concentrations above 100 mg N-NO_2.L^{-1} (Strous et al. 1999).

On the other hand, by keeping the initial ammonium concentration constant at 42 mg.L^{-1} in different batches, Egli et al. (2001) observed complete inactivation of the process at values above 185 mg N-NO_2.L^{-1}. Using granular biomass, improved nitrite tolerance was observed, with the IC_{50} at 350–400 mg N-NO_2.L^{-1} (Dapena-Mora et al. 2007; Cho et al. 2010; Hassan 2018), probably due to the limitation of substrate diffusion within the granules.

Other authors have used gel supports on which anammox bacteria are trapped. Batch experiments have shown that when these supports are immersed in solutions with concentrations higher than 274 mg N-NO_2.L^{-1}, anammox activity decreases. Finally, a 90% drop in anammox activity was observed when the nitrite concentration reached 750 mg N-NO_2.L^{-1} over a period of seven days in a bioreactor with continuous supply (Kimura et al. 2010).

1.8.12. *Ammonium ion (NH_4^+)*

In most wastewater treatment plants, the ammonium ion is one of the main forms of nitrogen. The literature reports that the anammox process is inhibited only at ammonium concentrations above 1,000 mg N.L^{-1} (Strous et al. 1999), but it is accepted that it is NH_3 rather than NH_4^+ that is the real inhibitor. Similarly, in short-term trials (batch tests), the median NH_3 inhibitory concentration is 38 mg N.L^{-1}, whereas in long-term tests in sequential batch reactors, process instability is observed beyond 20–25 mg-N.L^{-1} of ammonia (Dapena-Mora et al. 2007).

1.8.13. *Inhibition mechanisms and pH*

Ammonium does not easily diffuse across the lipid membrane of bacterial cells, which is not the case for ammonia (Kadam and Boone 1996). Thus, there is a chemical equilibrium between the molecule and the ion that can be affected by the pH (Mosquera-Corral et al. 2005):

NH_3/NH_4^+

$$NH_4^+ + OH^- \leftrightarrow NH_3 + H_2O,$$

The diffusion of ammonia into the cell changes the intracellular pH and neutralizes the transmembrane potential, which can cause cell death (Martinelle and Westlund 1996).

Similarly, the inhibition of the anammox process by nitrite is not caused by the nitrite itself but by nitrous acid (Fernández et al. 2012), and pH can also affect the chemical balance:

HNO_2/NO_2^-

$$HNO_2 + H_2O \leftrightarrow NO_2^- + H_3O^+$$

Thus, with a pKa of 9.24 and 3.3 for the acid–base pairs NH_3/NH_4^+ and HNO_2/NO_2^-, the pH of the anammox process is adjusted between 6.5 and 8.8, with an optimal pH of 8 (Oshiki et al. 2011). The pH must therefore be rigorously controlled during the operation of the process (Mosquera-Corral et al. 2005; Hassan 2018).

1.8.14. *Implementation using anammox bacteria*

Deammonification is a biological treatment process used to convert ammonia into nitrogen gas. It includes both nitritation and anaerobic ammonia oxidation processes (anammox).

Nitritation, the aerobic oxidation of ammonia N ($N\text{-}NH_4$) in nitrite-nitrogen ($N\text{-}NO_2^-$), is carried out by aerobic autotrophic bacteria that oxidize ammonia (AerAOB), a well-known process in wastewater treatment that constitutes the initial stage of biological nitrification of $N\text{-}NH_4$ into nitrates ($N\text{-}NO_3^-$).

These aerobic oxidizing bacteria are nitrifying bacteria, chemolithotrophic microorganisms, and include species of the genera *Nitrosomonas* and *Nitrosococcus*. They obtain their energy through the oxidation of nitrogenous compounds. More specifically, AOB obtain their energy by oxidizing ammonia/ammonium to nitrite.

AOB have a complex internal membrane system. Indeed, this is the location of key enzymes, such as ammonia monooxygenase, which oxidizes ammonia to hydroxylamine, or hydroxylamine oxidoreductase, which oxidizes hydroxylamine to nitric oxide. The latter is oxidized to nitrite by an unidentified enzyme (Kuypers et al. 2018). AOB produce nitrites according to the reaction shown below (Graham and Jolis 2017).

AOB and anammox bacteria can be combined to convert ammonia into nitrogen gas using a process called partial nitritation-anammox (PNA), where the AOB convert the available ammonia into nitrites. Then, anammox converts the NO_2^- and remaining $NH4^+$ into nitrogen gas (N_2), using NO_2^- as an electron acceptor:

$$NH_4^+ + 2.4\ O_2 + 6.72\ HCO_3^- \rightarrow 2.51\ NO_2^- + 1.04\ H_2O + 0.12\ C_5H_7NO_2 + 6.51\ H_2CO_3$$

In PNA, approximately 50% of the ammonium ions are initially converted to nitrites under aerobic conditions by AOB. In a second step, the remaining ammonium and nitrite ions are converted to nitrogen gas and a small fraction of nitrates by anammox bacteria (Graham and Jolis 2017):

$$2NH_4^+ + 0.34\ NO_2^- + 0.13\ H^+ + 1.38\ O_2 + 2.05\ HCO_3^- \rightarrow 1.02\ NO_2^- + 0.26\ NO_3^- + 3.09\ H_2O + 0.032\ C_5H_7NO_2 + 1.89\ H_2CO_3$$

The PNA process is commonly used in secondary wastewater treatment. In systems engineering, PNA can be found as a one- or two-stage application. The single-stage solution refers to carrying out the PNA process in a single reactor, while two-stage solutions maintain aerobic and anaerobic conditions in separate, but still connected, tanks.

The anammox reaction is carried out by anaerobic, autotrophic *Planctomycetales*, which are NH_4^+-oxidizing bacteria (AnAOB) that use NO_2^- as an electron acceptor to anaerobically oxidize the NH_4 into N_2 (Strous et al. 1999). Throughout the process, approximately 89% of the nitrogen is inorganic (NH_4 and NO_2) and ends up as N_2 gas and approximately 11% in the form of NO_3^-.

In the conventional denitrification process, NH_4^+ is first oxidized into NO_2^-, then into NO_3^- by autotrophic bacteria. Then, NO_3^- is biologically reduced into N_2 by heterotrophic bacteria (denitrification) with organic substrate consumption under anoxic conditions. Aerobic biological nitrification is well known and is carried out by the bacteria *Nitrosomonas* for the oxidation of NH_4^+ into NO_2^- and by *Nitrobacters* for the oxidation of NO_2^- into NO_3^-.

In the nitritation/denitritation process, the NH_4^+ is oxidized into NO_2^-, then biologically reduced to N_2 by heterotrophic bacteria consuming organic matter substrates under anoxic conditions (see Figure 1.21).

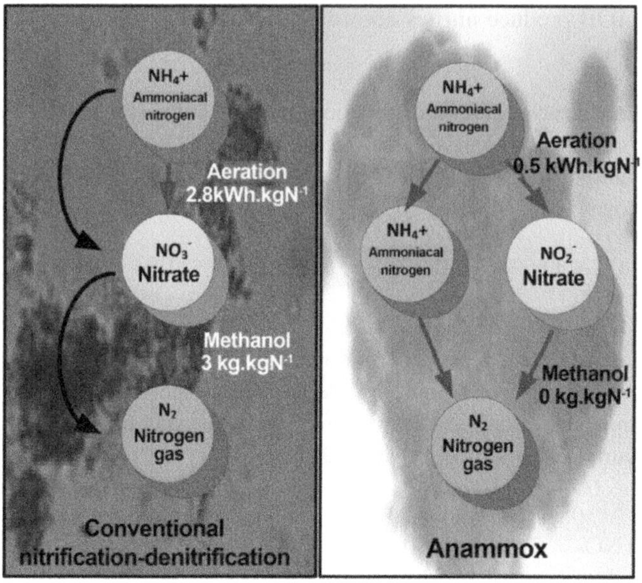

Figure 1.21. *Energy consumed by the conventional and anammox processes*

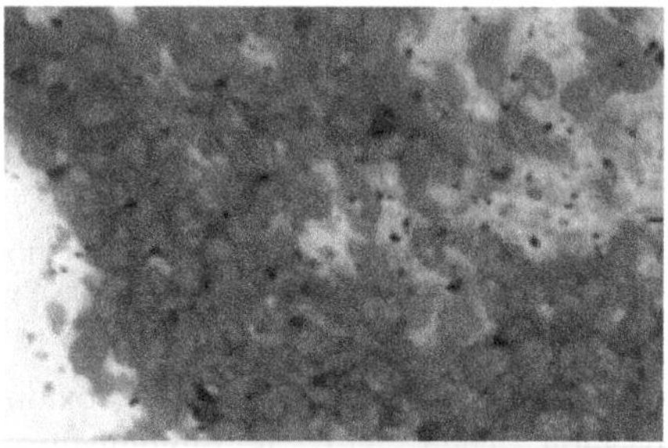

Figure 1.22. *Anammox bacteria in granules*

In the deammonification process, aerobic nitritation of NH_4^+ into NO_2^- represents the transformation of approximately 55% of the NH_4^+ of the influent. The remaining NH_4^+ is oxidized anaerobically with NO_2^- and into N_2 in the anammox process.

In practice, the process avoids the need for complete oxidation of $NO_2^-$and NO_3^-, followed by denitrification of NO_3^- into nitrogen gas.

Compared to conventional nitrification and denitrification, the demands for aeration and a carbon source are reduced by more than 50% and 100%, respectively.

It should be noted that biomass production is higher in the conventional process than in deammonification (20 g of COD biomass versus 3 g of COD biomass per mole of removed NH_4^+).

In the absence of dissolved oxygen, heterotrophic bacteria can oxidize the organic substrate with NO_2^- or NO_3^- as electron acceptors to reduce oxidized nitrogen to N_2 gas.

The reduction reactions of NO_2^- and NO_3^- are called denitritation and denitratation. The stoichiometries of biological denitrification and denitrification reactions with acetate consumption and heterotrophic biomass growth are as follows:

– Denitritation:

$$1.0.\ NO_2^- + 1.0\ H^+ + 0.24\ NH_4^+ + 0.975\ CH_3COO^- \rightarrow 0.5\ N_2 + 0.24\ C_5H_7O_2N + 0.015\ CO_2 + 0.735\ HCO_3^- + 1.235\ H_2O$$

– Denitratation:

$$1.0.\ NO_3^- + 1.0\ H^+ + 0.33\ NH_4^+ + 1.45\ CH_3COO^- \rightarrow 0.5\ N_2 + 0.33\ C_5H_7O_2N + 0.13\ CO_2 + 1.12\ HCO_3^- + 1.62\ H_2O$$

Therefore, 6.6 g of COD (acetate) is required per g of N-NO_3; for the denitritation of N-NO_2, just under 30%, that is, 4.5 g COD (acetate) per g de N-NO_2.

Anammox involves exoenergetic reactions for the oxidation of N-NH_4 into N-NO_2 and the absorption of CO_2 and nutrients through the growth of autotrophic anammox bacteria. The overall cellular synthesis reaction is written (Strous et al. 1999):

$$1.0.\ NH_4^+ + 1.32\ NO_2^- + 0.066\ HCO_3^- + 0.13\ H^+ \rightarrow 1.02\ N_2 + 0.26\ NO_3^- + 0.066\ CH_2O_{0.5}N_{0.15} + 2.03\ H_2O$$

During the anaerobic oxidation of ammonia, some nitrates are formed from nitrites, which can provide reducing power for the fixation of carbon dioxide (Schmidt et al. 2002). The removal of 1 mole of NH_4 requires 1.32 moles of NO_2 and produces 0.26 moles of NO_3.

The combination of nitritation and deammonification reactions leads to an overall reaction:

$$1.0.\ NH_4^+ + 0.804\ O_2 + 0.071\ HCO_3^- \rightarrow 0.436\ N_2 + 0.111\ NO_3^- + 0.009\ C_5H_7O_2N + 0.028\ CH_2O_{0.5}N_{0.15} + 1.038\ H^+ + 1.46\ H_2O$$

The stoichiometric ratio between nitrate production and ammonium consumption for anammox bacteria is 1:0.382. For field deammonification reactions, ammonia reductions of 90 to 95% and total nitrogen reductions of 80 to 85% (WERF 2014) have been obtained.

Table 1.12 summarizes the advantages of the deammonification process compared to the conventional denitrification process.

Parameters	Deammonification	Nitritation – denitritation	Nitrification – denitrification
O_2 demand	1.84 $gO_2.gN\text{-}NH_4^{-1}$ removed	2.65 $gO_2.gN\text{-}NH_4^{-1}$ removed	3.3 $gO_2.gN\text{-}NH_4^{-1}$ removed
COD demand (acetate)	0	4.5 g acétate $COD.gN\text{-}NO_2$ removed	6.6 g acetate $COD.gN\text{-}NO_2$ removed
Biomass production	0.12 g MVS $gN\text{-}NH_4^{-1}$ removed	1.5 g MVS $gN\text{-}NH_4^{-1}$ removed	1.93 g MVS $gN\text{-}NH_4^{-1}$ removed

Table 1.12. *Comparison of conventional deammonification and denitrification processes (WERF 2014)*

1.8.14.1. *Influence of the C/N ratio*

Theoretically, 2.86 g of COD would be needed to eliminate 1 g of $N\text{-}NO_3^-$ (Hellinga et al. 1999; Daigger 2014). However, since a significant amount of COD would be oxidized during the aeration phase, the C/N ratios actually required for complete nitrogen removal are higher (5 to 7) (Hellinga et al. 1999; Sahinkaya et al. 2014). C/N ratios of 1.94 to 3 would be required for the removal of nitrogen via the NO_2^- pathway (nitritation–denitritation). This nitrogen removal option saves approximately 40% of COD and 25% of aeration costs compared to nitrification/denitrification since the NH_4^+ is only oxidized into NO_2^- as opposed to NO_3^- in the case of classic nitrification/denitrification (Hellinga et al. 1999; Van Kempen et al. 2001; Noutsopoulos et al. 2018; Dobbeleers et al. 2020).

Coupling nitritation with denitrification requires control of parameters such as pH, OD, temperature and the sludge retention time (SRT)/hydraulic retention time (HRT) ratio or even G/t (sludge age/HRT). This is done to suppress NOB growth while promoting AOB growth. Generally, biological reactors operate at a temperature of 30–40°C and a variable pH between 7 and 8.

1.8.14.2. *Influence of temperature*

Between 5 and 20°C, NO_2^- oxidizes more quickly than NH_4^+, leading to the generation of NO_3^-. However, at higher temperatures (30–40°C), the reverse is also true. For example, the difference in growth rates is the basis of the Sharon process, which seeks to limit nitration while promoting nitration. During the process, a combination of high temperatures (30–40°C) and the reaction time (1 day) leads to the leaching of NOB. Essentially, the high growth rates of AOB compensate for the biomass removed from the system.

1.8.14.3. *Inhibition of anammox bacteria reactions*

Furthermore, high temperatures and high concentrations of NH_4^+ and NO_2^- also limit NOB activity because they respectively dissociate into free ammonia (FA) and free nitrous acid (FNA), both of which are more toxic to NOBs than to AOBs (Anthonisen et al. 1976; Duan et al. 2021).

The inhibitory concentrations of FA and FNA for NOB are 0.10 to 1 $mg.L^{-1}$ and 0.011 to 0.070 $mg.L^{-1}$, while those for AOB are 10 to 150 $mg.L^{-1}$ and approximately 0.40, respectively (Anthonisen et al. 1976; Chen et al. 2020).

1.8.14.4. *Influence of pH*

To completely remove nitrogen from wastewater by nitritation–denitritation, a COD supplement is required for low-COD wastewater. The process could be implemented in a continuously stirred system in which a COD source is added during a non-aerated phase (Lemaire et al. 2008). Since aeration lowers the pH and denitrification raises it, this parameter is used to control the aeration and dosing of the COD sources. This strategy is beneficial because the pH is adjusted during nitritation and denitrification, thus eliminating the need to add an alkali (Hellinga et al. 1998). If the incoming wastewater contains high C/N ratios (> 2), then the same wastewater could be used as a COD source by implementing a phased feeding strategy, as Lemaire et al. (2003) demonstrated. Other process control methods include sequencing operations to define time intervals. Strategies based on the oxidation–reduction potential (ORP) could also be used to regulate the process, as previously demonstrated (Claros et al. 2012).

1.8.15. *Nitrogen removal processes using anammox*

Wastewater treatment plant outlet standards comply with European standards. These standards, based on the quality of the receiving environment, most often require nitrogen treatment in addition to organic matter (COD and BOD) and particulate matter.

Ammoniacal nitrogen (NH_4) is the main nitrogen compound contained in the treatment plant's inlet water and must be treated in situ.

With the digestion process and the combination of "thermal hydrolysis + digestion", the effluents generated during the digested sludge dewatering stage return to the start containing high concentrations of ammoniacal nitrogen but very little organic carbon (transformed into methane in the digester).

Although representing only 1 to 2% of the plant's inlet flow, these return flows can contain 15 to 30% of the nitrogen load entering the plant. Table 1.13 illustrates the impact of these return flows from the dewatering plant according to the treatment process.

Sludge treatment process	**Concentration N-NH_4^+ (mg.L^{-1})**	**Nitrogen load recycled**
Digestion alone	500–1,500	≈ 10–15%
Digestion with thermal hydrolysis	2,000–4,000	≈ 25–30%

Table 1.13. *NH_4 concentrations for the returning loads from the digester*

This nitrogen overload results in a significant increase in the cost of biologically treating the water supply by generating:

– an additional energy cost linked to the increased oxygen requirements for processing nitrogen overloads during the nitrification phase;

– an additional cost for reagents related to the potential need for a carbon substrate (methanol) during the denitrification phase;

– a risk of nitrogen non-compliance when the size and variability of return loads are not correctly taken into account when setting the initial size of the water piping.

Thus, separate treatment of digested recirculated water upstream of the water line is paramount if we wish to avoid the consequences linked to implementing boosted digesters (external inputs, prior thermal hydrolysis of sludge, etc.).

Several technical solutions have been developed to implement the deammonification process. Table 1.14 presents the technical solutions, which differ in the methods for growing and preserving anammox bacteria, the number of stages, the configuration and the control strategies. These include fluidized bed reactors, SBRs, moving bed reactors (MBBRs) and rotating biological contactors.

Process	**Company**
DEMON®	APG Neuros (USA)
Anita™ Mox	Veolia (France)
ANAMMOX®	Paques (the Netherlands)
DeAmmon	Purac/Läckeby AB (Sweden)
Terra-N	Clariant/SÜD-Chemie AG (Germany)
Sharon®	Université (the Netherlands)

Table 1.14. *Processes using anammox bacteria*

1.8.15.1. *DEMON® process*

The DEMON® process, acronym for "deammonification", involves two stages: the partial nitritation of ammonia and the subsequent anaerobic oxidation of residual ammonia by nitrites into nitrogen gas. Partial nitritation requires only 40% oxygen compared to conventional nitrification. No carbon (methanol) dosing is necessary due to the autotrophic nature of the process.

The DEMON® process is an SBR process with aerated/anaerobic conditions controlled by supply, aeration and mixing mini-cycles in non-aerated conditions that maintain aerobic/anaerobic conditions. The aeration periods promote the oxidation of some of the NH_4^+ into NO_2^-. Cycle times depend on the concentration of NH_4^+. When these concentrations are high, the cycle times are longer.

Controlling dissolved oxygen is important, and it must be maintained at values below 0.6 mg $O_2.L^{-1}$. Such control is important to avoid the inhibition of anammox due to nitrite accumulation, and also to ensure the survival of NOB.

The enormous impact on resource consumption demonstrates the nitrogen cycle. Denitrification and nitration do not occur; nitritation only partially occurs. Total nitrogen removal is achieved using only a very small amount of oxygen.

Figure 1.23. *DEMON® process*

1.8.15.2. *Anita™ Mox process*

A high ammonia concentration and a low chemical oxygen demand (COD) are characteristic of the water discharged from the dewatering of digested sludge. This ammonia load is generally returned to the main wastewater stream. However, its treatment by conventional nitrification/denitrification proves very costly or, in some cases, can disrupt downstream sludge digestion processes aimed at reducing sludge volume or improving biogas production.

Reducing sludge volumes while producing energy (conventional anaerobic digestion, thermal reduction processes) generates effluents. This recirculated water, heavily laden with NH_4^+, is often sent back to the start of the treatment line and consequently increases the nitrogen load of the water system by 15 to 20%.

This has a direct impact on the capacity of aeration systems, the size of biological tanks and the quality of effluents. These effluents can exceed the standards set for nitrogen without resorting to a post-denitrification step, which involves the costly injection of methanol as a carbon substrate.

To remedy this, AnoxKaldnes™ (Veolia) developed Anita™ Mox, an alternative to conventional nitrification/denitrification processes that limits the impact of digestion on the main treatment line of the wastewater treatment plant. Anita™ Mox is another application of a well-established technology, the mobile bed biofilm reactor (MBBR) or IFAS. It combines a single-stage MBBR with a proven aeration control strategy. Anita™ Mox significantly reduces the ammonia load (> 85–90%) and eliminates 75 to 85% of the nitrogen in the main treatment process, making it an ideal solution for modernizing an existing, overloaded wastewater treatment plant at a competitive cost. An anammox process comprises two stages: aerobic nitritation by ammonia-oxidizing bacteria and anoxic oxidation of ammonia by anammox bacteria. In Anita™ Mox, both steps take place simultaneously in a single reactor and both aerobic and anoxic reactions occur in a single MBBR equipped with plastic supports specifically designed to retain biofilm. Specific pH, temperature and oxygen level conditions are maintained in the reactor; these provide ammonia-removing bacteria with an optimal environment, allowing them to develop as biofilms on the substrates and preventing their leaching.

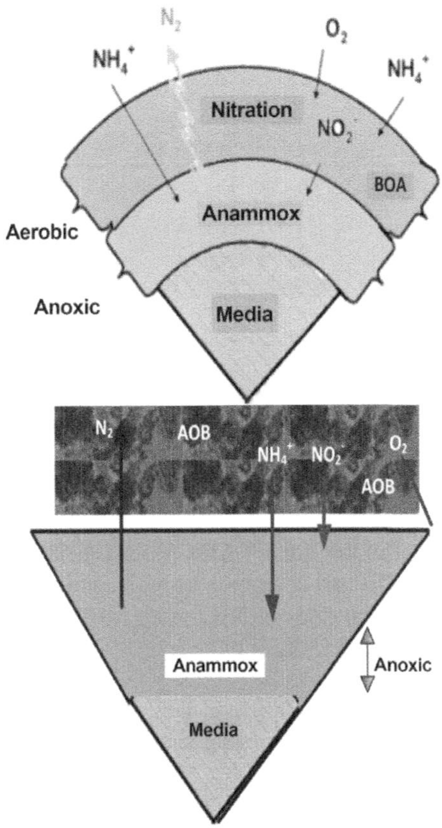

Figure 1.24. *Anita™ Mox principle: a) MBBR; b) IFAS*

On the substrate biofilm, aerobic and anaerobic zones develop simultaneously. AOB develop on the outer layer, exposed to the NH_4^+ concentration and dissolved oxygen. Anammox develops on the inner layer close to the substrate (see Figure 1.24).

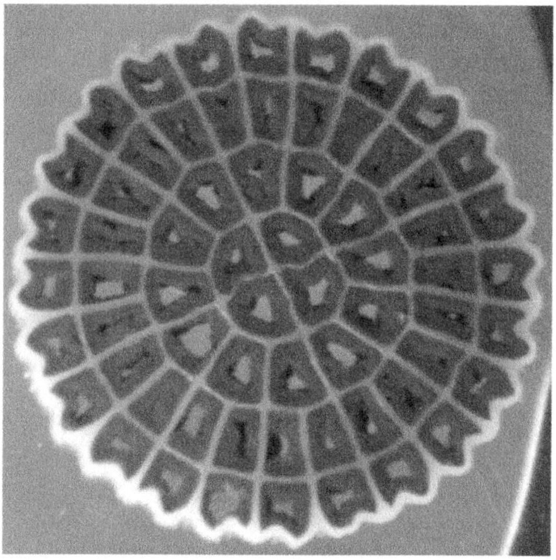

Figure 1.25. *Biofilm on AnoxKaldnes™ substrate*

The IFAS process comprises two stages during which bacteria develop in aerobic and anaerobic zones. AOB bacteria are present in free cultures (activated sludge), while anammox bacteria are present on the biofilm.

The operating parameter of the Anita™ Mox process is the O_2 concentration, which influences the activity of AOB and anammox bacteria, and therefore the efficiency of deammonification. The removal rates of NH_4^+ with MBBR amount to 1.2 kg $N.m^{-3}$, whereas in IFAS mode, they reach 3 kg $N.m^{-3}$.

The process has opened up attractive new solutions for nitrogen removal applications due to the ease of modernizing the IFAS process in activated sludge systems with existing clarifiers.

In 2010, after two years of pilot tests in France, Denmark and Sweden, the Anita™ Mox process was implemented on an industrial scale at the Sjölunda wastewater treatment plant near Malmö in Sweden (550,000 EH). Today, it processes 200 kg of nitrogen per day from sludge digestion. This unit also serves as a "biofarm" to produce substrates with a mature and efficient biofilm.

This strategy was implemented to accelerate the start-up of other Anita™ Mox industrial units at the wastewater treatment plants of Sundets in Växjö, Sweden; Holbaek and Grindsted in Denmark; and even recently other facilities in the United States.

In France, the recent development of anaerobic sludge digestion to produce biogas and move towards energy self-sufficiency makes the treatment of nitrogenous wastewater through these new biological processes increasingly acceptable and feasible.

Anita™ Mox can also be applied to other effluents high in NH_4 and with little BOD, such as landfill leachate, industrial effluents or leachate from UASB-type anaerobic processes.

Essentially, Anita™ Mox is a process dedicated to the treatment of NH_4 in return supply lines, simultaneously implementing:

– the aerobic nitritation step ($NH_4 \rightarrow NO_2$; nitrifying bacteria);

– the anoxic deammonification stage without a carbon source ($NH_4 + NO_2 \rightarrow N_2$; anammox bacteria).

Compared to conventional nitrification/denitrification, the process allows:

– a 60% reduction in O_2 requirements;

– the complete removal of the need for carbon sources;

– an 80% reduction in sludge production.

The Anita™ Mox process was adapted into the synergistic interaction of this type of biofilm, composed primarily of anammox bacteria, with free biomass (conventional activated sludge) within the MBBR. This configuration requires the implementation of a clarifier downstream of the MBBR to recirculate the activated sludge back into the reactor. This is called the Integrated Fixed-Film Activated Sludge (IFAS) configuration. This difference in the spatial organization of bacteria allows the Anita™ Mox process in the IFAS configuration to be two to three times more compact than the MBBR configuration despite the addition of equipment such as the clarifier.

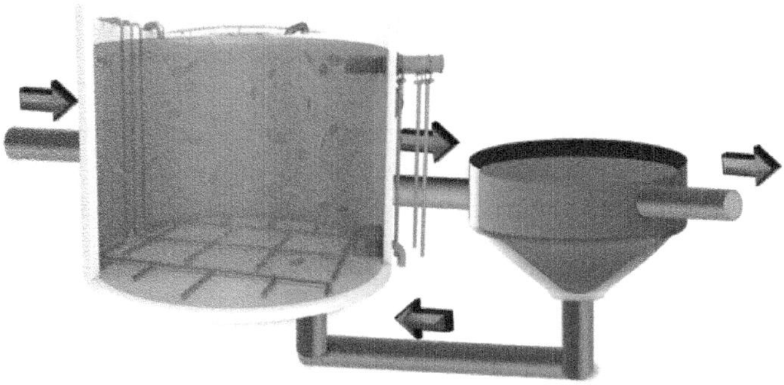

Figure 1.26. *Diagram of the Anita™ Mox process in IFAS configuration with its clarifier*

Site	Country	Capacity	Type of effluent	Date
Sjölunda (Malmö)	Sweden	200 kg $N.d^{-1}$	Digestion	2010
Sundets (Växjö)	Sweden	430 kg $N.d^{-1}$	Digestion	2012
Holbaek	Denmark	120 kg $N.d^{-1}$	Digestion	2012
Grindsted	Denmark	110 kg $N.d^{-1}$	Digestion	2013
James River (Newport)	United States	250 kg $N.d^{-1}$	Digestion	2013
South Durham	United States	330 kg $N.d^{-1}$	Digestion	2014
Egan (Chicago)	United States	940 kg $N.d^{-1}$	Digestion	2016
Mars Candy	Poland	340 kg $N.d^{-1}$	F&B return after AnMBR	2015
Starch	Germany	300 kg $N.d^{-1}$	F&B return after UASB	2016
Viikinmäki (Helsinki)	Finland	320 kg $N.d^{-1}$	Municipal sidestream	2016
Sobacken (Boraas)	Sweden	800 kg $N.d^{-1}$	Municipal sidestream	2017
Denver Metro	United States	3,000 kg $N.d^{-1}$	Municipal sidestream	2018
Tomahawk (Johnson)	United States	–	Municipal sidestream	2018

Table 1.15. *Anita™ Mox (Veolia) references*

1.8.15.3. *Paques process*

Since Paques launched the first biological nitrogen filtration system in 2002, ANAMMOX® has become the most widely used nitrogen removal system in the world; it offers a treatment capacity of over 160,000 kg of nitrogen per day. The ANAMMOX® conversion is an elegant shortcut in the natural nitrogen cycle.

Anammox bacteria convert ammonium (NH_4^+) and nitrite (NO_2^-) into nitrogen (N_2) gas, which is environmentally neutral. Paques's patented ANAMMOX® technology combines the anammox process with partial nitritation (conversion of ammonium to nitrite). The ANAMMOX® process works with naturally cultivated concentrated granular sludge, creating optimal conditions for microbial synergy.

1.8.15.4. *Sharon® process*

Initially developed by a group of researchers at Delft University in the Netherlands, the Sharon® process is a two-step process. Its acronym is derived from the overall name: Single-reactor System for High Ammonia Removal Over Nitrite. The Sharon® process is a biological system for treating effluents with high ammoniacal nitrogen content. It is based on the selection of a specific microorganism culture, with conditions defined by various process parameters such as sludge age, pH and temperature, in order to implement so-called partial nitrification or a nitrate shunt. The aim is to treat ammonia nitrogen with aerobic nitritation ($N\text{-}NH_4 \rightarrow NO_2^-$) followed by anaerobic denitrification ($NO_2^- \rightarrow N_2$), the classic nitrate stage ($NO_2^- \rightarrow NO_3^-$). The treatment system includes an anoxic tank for denitrification (non-aerated but continuously agitated) and an aerated and agitated tank for nitritation located outside the former. The stored effluent (at 30–40°C) temporarily held in the buffer tank is cooled in a heat exchanger and then fed into the anoxic reactor. The sludge–effluent mixture then passes through the aerated reactor via a wall pump and returns to the anoxic tank by overflow. The hydraulic retention time of the system is identical to the sludge age, as the sludge neither settles nor recirculates. For process optimization, the tank pH can be regulated by adding sodium hydroxide. Phosphorus rebalancing is achieved by adding phosphoric acid, and organic carbon is supplied in the form of methanol for denitrification. According to the manufacturer, the Sharon® process and its nitrate shunt should help reduce energy costs since the O_2 consumption is 25% lower during nitrification and 40% for carbon during denitrification. It should reduce biological sludge production by 25% by decreasing the number of steps in the chemical reactions. Finally, the denitrification rate is also 1.5 to 2 times higher with partial nitrification compared to conventional nitrification.

Over the past decade, an increasing number of researchers have begun to focus on the traditional anammox process. However, the traditional anammox process faces more practical challenges.

Firstly, there is little focus on NH_4^+ in municipal wastewater; it is generally 20 to 60 mg-N L^{-1} (Gilbert et al. 2014). The lack of substrate inhibits the activity of ammonia-oxidizing bacteria and anammox bacteria, and the lower concentration of free ammonia and free nitrous acid makes it more difficult to inhibit the activity of NOB.

Secondly, the temperature of municipal wastewater is low in winter (generally < 15°C), which considerably inhibits the activity of anammox bacteria (Gilbert et al. 2014). Furthermore, the GDP growth rate is faster than that of AOB at low temperatures, leading to substrate competition between anammox and NOB.

Thirdly, the carbon-to-nitrogen (C/N) ratio in the main influent is higher. A high C/N ratio promotes the proliferation of heterotrophic bacteria, intensifying competition between heterotrophic and anammox bacteria for the NO_2^-, reducing the overall nitrogen removal efficiency.

Most importantly, the slow growth rate of anammox bacteria and the interference of complex microbial communities have become two bottlenecks with the primary anammox process. Indeed, the growth rate of Anammox bacteria is slow. It is only 0.002 to 0.007 h^{-1}, while the maximum growth rates representative of AOB, NOB and denitrifiers are respectively 0.09 h^{-1}, 0.06 h^{-1} and 0.26 h^{-1}. Consequently, this slow growth rate results in a long start-up time and a low nitrogen removal rate. The anammox process requires sufficient nitrite acting as an electron acceptor. However, nitrogen in municipal wastewater exists primarily in the form of NH_4^+; it is generally necessary to combine partial nitrification with anammox. First, ammonia is oxidized to nitrite by AOB bacteria. Then, the remaining ammonia is oxidized to nitrogen gas by the accumulated nitrite through the action of anammox bacteria.

1.8.16. *Applications*

A large number of configurations incorporating anammox bacteria have been proposed in treatment processes. Figures 1.27 to 1.29 present three of them.

1.8.16.1. *Mainstream configuration*

The anammox system is installed in a continuous configuration, on the same line as the main treatment facilities. Conventional treatment removes organic carbon. Nitritation is then carried out in conjunction with the anammox system.

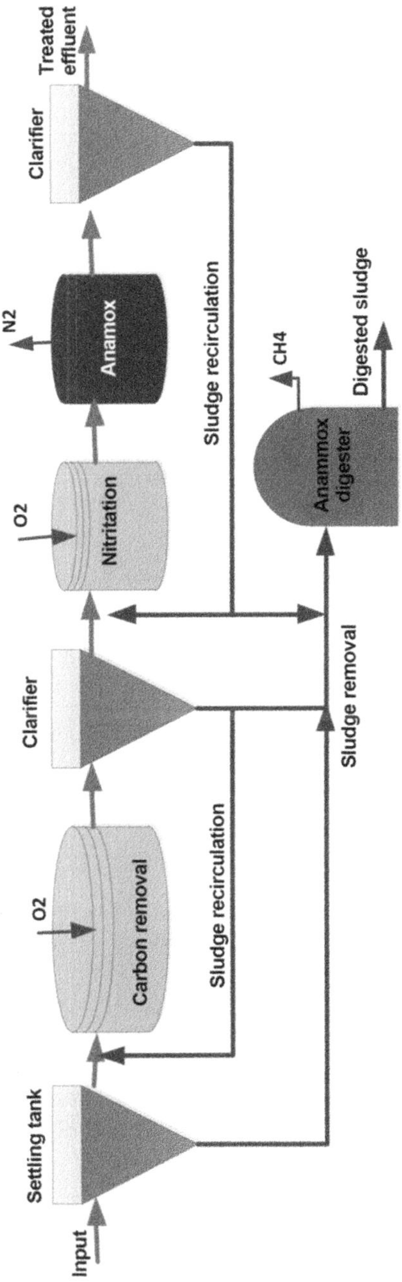

Figure 1.27. *Anammox installed in the main process*

1.8.16.2. *Sidestream configuration*

Anammox treats high NH_4 concentrations present in the digester supernatant rich in inlet effluent or cells from the DN.

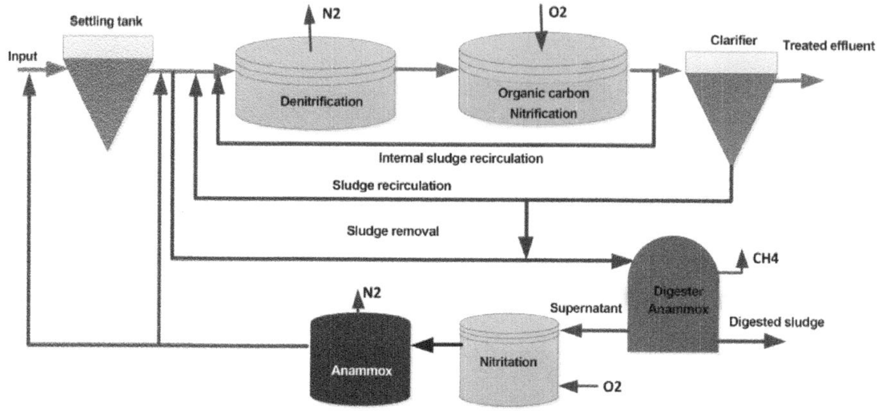

Figure 1.28. *Anammox installed on the secondary line*

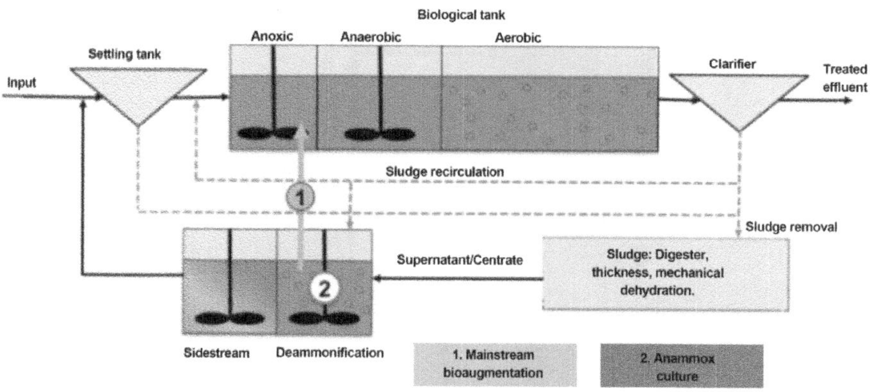

Figure 1.29. *Mainstream and sidestream configurations for part of returning sludge*

Figure 1.29 shows the solution that prevents potential failure of the anammox reactor (second stage). Lateral deammonification can be used as a solution for nitrogen treatment, as well as for cultivating anammox for bio-augmentation of the biological tank. A rapid increase in the inlet nitrogen load (i.e. due to a failure of the sidestream deammonification process) can lead to a decrease in the quality of the main reactor effluent, particularly at cold temperatures. The additional AOB mass increases the nitrogen removal capacity of the main reactor and protects against effluent quality degradation during a period of increased load or a sudden loss of biomass.

1.9. MABR: a new opportunity

Like activated sludge, wastewater treatment in a membrane aerated biofilm reactor (MABR) uses microorganisms to nitrify and denitrify wastewater, but the MABR uses a spirally wound, self-aspirating membrane with a large surface area, which passively supplies oxygen instead of manually pumping air into the water. This configuration offers several advantages. Aeration is carried out at a pressure close to atmospheric pressure, virtually eliminating the costly air compression that accounts for 40–60% of energy consumption in existing processes. A natural, resistant layer of microorganisms forms on the oxygen-rich surface of the membrane, where bacteria can use oxygen most efficiently. In contrast, with conventional bubble aeration, most air bubbles rise to the surface before microorganisms can use them (Ni et al. 2016).

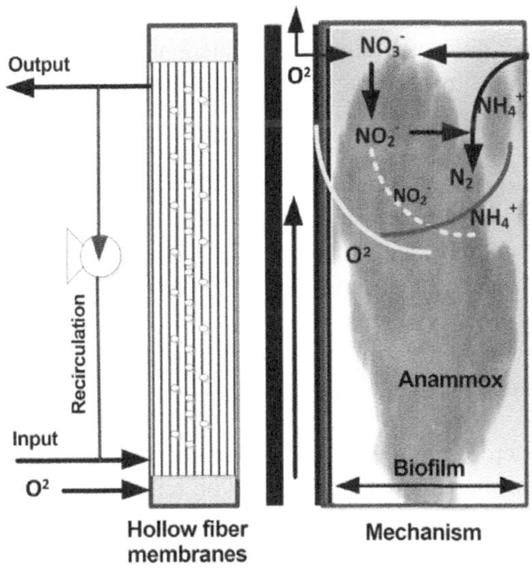

Figure 1.30. *Basic MABR operating principle diagram*

Conditions in the rest of the MABR tank remain anoxic, so anaerobic denitrifying bacteria can do their work in the same reservoir as aerobic bacteria. This simultaneous nitrification–denitrification (SND) process allows for a smaller footprint compared to traditional processes that require multiple chambers.

1.10. Conclusion

Several low-carbon nitrogen removal processes have been developed, including nitritation–denitrification, anammox-regulated processes, bioelectrochemical processes, biofilm-on-membrane association with O_2 diffusion and autotrophic denitrification.

Identifying which biological processes are best for nitrogen removal in municipal wastewater treatment is important for future development, as there are various emerging processes that can have a significant impact on resource use and climate change. In theory, processes using anammox surpass all others in terms of net energy production while significantly reducing sludge treatment requirements. However, the efficiency and reliability of anammox treatment have not yet been demonstrated, even though the Veolia MBBR and IFAS installations perform very well. Certainly, the difficulties encountered include the slow growth of organisms, the effects of temperature and the challenges of balancing oxygen supply with the significant diurnal variations in nitrogen that occur in the main flow. Insufficient oxygen leads to low ammonia removal efficiency, while excessive oxygen results in excessive nitrate production. However, given the significant potential environmental benefits of this novel combination of processes, research aimed at accelerating large-scale development appears to be an urgent need.

Bioelectrochemical processes are limited by the low wastewater conductivities, and low applied loads restrict their large-scale application. Autotrophic denitrification induced by sulfur and its compounds is only viable for moderate removed loads (1 $kg\text{-}N.m^{-3}\ d^{-1}$). However, it generates harmful chemicals that require downstream disposal to avoid negative environmental impacts. Overall, nitrogen removal processes using anammox offer the best alternatives to nitrification/denitrification in terms of COD demand, simplicity and process performance.

There are many aerobic and anaerobic biological processes and different treatment schemes for transforming organic matter and/or removing nitrogen from domestic wastewater. Significant reductions in oxygen requirements and no need for organic matter for nitrogen reduction are often cited as advantages of using the newer anammox organism approach for nitrogen removal rather than the traditional nitrification/denitrification method, the most common one used today. However, treatment schemes differ, and for some, the suggested benefits may not be acceptable. When nitrification/denitrification is used, an anoxic tank is commonly used at the beginning of the process, and the nitrates formed by the nitrification stage are then recycled to the anoxic tank, which benefits from the presence of organic matter in untreated wastewater.

This significantly reduces oxygen requirements and the need to add organic matter (methanol) when denitrification is carried out at the end of the process.

So, when are such statements correct and when are they not? What factors in the wastewater composition, regulatory requirements and treatment scheme alter the most appropriate treatment process? To help make such judgments in different circumstances, the stoichiometry of the various biological processes involved and the different treatment approaches used must be determined and compared.

1.11. References

Abeysiriwardana-Arachchige, I.S.A., Munasinghe-Arachchige, S.P., Delanka-Pedige, H.M.K., Nirmalakhandan, N. (2020). Removal and recovery of nutrients from municipal sewage: Algal vs. conventional approaches. *Water Res.*, 175, 115709.

Ali, M., Chai, L.-Y., Tang, C.-J., Zheng, P., Min, X.B., Yang, Z.H., Xiong, L., Song, Y.X. (2013). The increasing interest of anammox research in China: Bacteria, process development, and application. *BioMed Res. Int.*, 56, 1–21.

Amatya, I.M., Kansakar, B., Tare, V., Fiksdal, L. (2005). Impact of temperature on biological denitrification process. *J. Instit. of Eng.*, 7(1), 1–6.

Anthonisen, A.C., Loehr, R.C., Prakasam, T., Srinath, E. (1976). Inhibition of nitrification by ammonia and nitrous acid. *J. Water Pollut. Control Fed.*, 48, 835–852.

Argaman, Y. (1994) Nitrogen control considerations in activated sludge systems design. *New World Water*, 24, 77–80.

Argaman, Y., Papkov, G., Ostfeld, A., Rubin, D. (1999). Single-sludge nitrogen removal model: Calibration and verification. *J. Environ. Eng.*, 125(7), 608–617.

Baalsrud, K. (1954). Studies on *Thiobacillus denitrificans*. *Arch. für Microbiol.*, 20(1), 34–62.

Batchelor, W. and Lawrence, A.W. (1978). Autotrophic denitrification using elemental sulfur. *J. Water Pollut. Control Fed.*, 50(8), 1986–2001.

Beijerinck, W.M. (1910). Bildung und Verbrauch von Stickoxydul durch Bakterien. *Zentralbl. Bakteriol. Parasitenkd. Infektionskr. Hyg. Abt.*, 25, 30–63.

Ben Khaled, J. (2016). Couplage de la dénitrification autotrophe sur soufre aux systèmes de traitement biologique des eaux usées domestiques applicables au secteur de l’assainissement décentralisé. PhD Thesis, Université du Québec, Quebec.

Bergaust, L., Mao, Y., Bakken, L.R., Frostegard, A. (2010). Denitrification response patterns during the transition to anoxic respiration and post transcriptional effects of suboptimal pH on nitrogen oxide reductase in *Paracoccus denitrificans*. *Appl. Environ. Microbiol.*, 76, 6387–6396.

Blais, J.F., Tyagi, R.D., Meunier, N., Auclair, J.C. (1994). The production of extracellular appendages during bacterial colonization of elemental sulphur. *Process Biochem.*, 29, 475–482.

Bowden, G., Stensel, H.D., Tsuchihashi, R. (2016). Technologies for sidestream nitrogen removal. Report 15, Water Environment Research Fondation.

Brandes, J., Devol, A.H., Deutsch, C. (2007). New developments in the marine nitrogen cycle. *Chemical Reviews*, 107(2), 577–689.

Brummer, I.H.M., Felske, A.D.M., Wagner-Dobler, I. (2004). Diversity and seasonal changes of uncultured Planctomycetales in river biofilms. *Appl. Environ. Microbiol.*, 70(9), 5094–5101.

Campos, J.L., Carvalho, S., Portela, R., Mosquera-Corral, A., Mendez, R. (2008). Kinetics of denitrification using sulphur compounds: Effects of S/N ratio, endogenous and exogenous compounds. *Bioresour. Technol.*, 99(5), 1293–1299.

Canler, J.-P., Perret, J.-M., Choubert, J.-M. (2007). Le traitement du carbone et de l'azote pour des stations d'épuration de type boue activée confrontées à des fortes variations de charge et à des basses températures. Report, FNDAE, Paris.

Cao, S. and Zhou, Y. (2019). New direction in biological nitrogen removal from industrial nitrate wastewater via anammox. *Appl. Microbiol. Biotechnol.*, 103, 7459–7466.

Cardoso, N., Guerra, F., Chaves, H.O.C. (2006). Influence of nitrogen in the production and quality of melon, *Revista Caatinga*, 19, 153–160.

Cecconet, D., Sabba, F., Devecseri, M., Callegari, A., Capodaglio, A.G. (2020). In situ ground-water remediation with bioelectrochemical systems: A critical review and future perspective. *Environ. Int.*, 137, 105550.

Chen, Y., Jiang, M., Zheng, X. (2020). Enhancement of denitrification performance with reduction of nitrite accumulation and N2O emission by Shewanella oneidensis MR-1 in microbial denitrifying process. *Water Res.*, 169, 115242–115253.

Cho, S., Takahashi, Y., Fujii, N., Yamada, Y., Satoh, H., Okabe, S. (2010). Nitrogen removal performance and microbial community analysis of an anaerobic upflow granular bed anammox reactor. *Chemosphere*, 78(9), 1129–1135.

Christensson, M., Ekstrom, S., Chan, A., Le Vaillant, E., Lemaire, R. (2013). Experience from start-ups of the first Anita Mox® plants. *Water Sci. Technol.*, 67, 2677–2684.

Claros, J., Serralta, J., Seco, A., Ferrer, J., Aguado, D. (2012). Real-time control strategy for nitrogen removal via nitrite in a Sharon® reactor using pH and ORP sensors. *Process Biochem.*, 47, 1510–1515.

Claus, G. and Kutzner, H.J. (1985). Physiology and kinetics of autotrophic denitrification by *Thiobacillus denitrificans*. *Appl. Microbiol. Biotechnol.*, 22(4), 283–288.

Constantine, T., Sandino, J., Houweling, D., Stephan, S., Yin, H., Nielsen, P. (2016). Incorporating leading edge mainstream deammonification into full-scale advanced BNR facilities. *Proc. Water Environ. Fed.*, 2016, 1007–1018.

Cui, Y.-X., Biswal, B.K., Van Loosdrecht, M.C.M., Chen, G.-H., Wu, D. (2019). Long term performance and dynamics of microbial biofilm communities performing sulfur-oxidizing auto-trophic denitrification in a moving-bed biofilm reactor. *Water Res.*, 166, 115038.

Daigger, G.T. (2014). Oxygen and carbon requirements for biological nitrogen removal processes accomplishing nitrification, nitritation, and anammox. *Water Environ. Res.*, 86, 204–209.

Daigger, G.T. and Littleton, H.X. (2014). Simultaneous biological nutrient removal: A state-of-the-art review. *Water Environ. Res.*, 86(3), 245–257.

Dalsgaard, T., Thamdrup, B., Canfield. D.E. (2005). Anaerobic ammonium oxidation (anammox) in the marine environment. *Research in Microbiology*, 156(4), 457–464.

Dapena-Mora, A., Fernández, I., Campos, J.L., Mosquera-Corral, A., Méndez, R., Jetten, M.S.M. (2007). Evaluation of activity and inhibition effects on anammox process by batch tests based on the nitrogen gas production. *Enzym. Microb. Technol.*, 40, 859–865.

Deronzier, G., Schétrite, S., Racault, Y., Canler, J.-P., Liénard, A., Héduit, A., Duchène, P. (2002). Traitement de l'azote dans les stations d'épuration biologique des petites collectivités. Report, FNDAE, Ministère de l'Agriculture de la Pêche et de l'Alimentation, CEMAGREF, Paris.

Di Capua, F., Pirozzi, F., Lens, P.N.L., Esposito, G. (2019). Electron donors for autotrophic denitrification. *Chem. Eng. J.*, 362, 922–937.

Dimitrova, I., Dabrowska, A., Ekström, S. (2020). Start-up of a full-scale partial nitritation-anammox MBBR without inoculum at Klagshamn WWTP. *Water Sci. Technol.*, 81, 2033–2042.

Dobbeleers, T., Caluwé, M., Dockx, L., Daens, D., D'aes, J., Dries, J. (2020). Biological nutrient removal from slaughterhouse wastewater via nitritation/denitritation using granular sludge: An on site pilot demonstration. *J. Chem. Technol. Biotechnol.*, 95, 111–122.

Dobbeleers, T., Dockx, L., Caluwé, M., Dries, J. (2022). Nitrous oxide formation during simultaneous phosphorus and nitrogen removal in aerobic granular sludge treating different carbon substrates. *Bioresour. Technol.*, 345, 125–136.

Dosta, J., Fernández, I., Vázquez-Padín, J.R., Mosquera-Corral, A., Campos, J.L., Mata Álvarez, J., Méndez, R. (2008). Short- and long-term effects of temperature on the anammox process. *J. Hazardous Mater.*, 154(1–3), 693–698.

Du, R., Cao, S., Li, B., Niu, M., Wang, S., Peng, Y. (2017). Performance and microbial community analysis of a novel DEAMOX based on partial-denitrification and anammox treating ammonia and nitrate wastewater. *Water Res.*, 108, 46–56.

Duan, H., Zhao, Y., Koch, K., Wells, G.F., Zheng, M., Yuan, Z., Ye, L. (2021). Insights into nitrous oxide mitigation strategies in wastewater treatment and challenges for wider implementation. *Environ. Sci. Technol.*, 55(11), 7208–7224.

Duchène, P. (1990). Elimination de l'azote dans les stations d'épuration des petites collectivités. Report, FNDAE, Paris.

Egli, K., Fanger, U., Alvarez, P.J., Siegrist, H., van der Meer, J.R., Zehnder, A.J. (2001). Enrichment and characterization of an anammox bacterium from a rotating biological contactor treating ammonium-rich leachate. *Arch. Microbiol.*, 175(3), 198–207.

Ekama, G.A. and Wentzel, M.C. (2008). Nitrogen removal. In *Biological Wastewater Treatment: Principles, Modelling and Design*, Henze, M., van Loosdrecht, M.C.M., Ekama, G.A., Brdjanovic, D. (eds). IWA Publishing, London, 87–138.

Essila, N.J., Semmens, M.J., Voller, V.R. (2000). Modeling biofilms on gas-permeable supports: Concentration and activity profiles. *J. Environ. Eng.*, 126(3), 250–257.

Fernández, I., Vázquez-Padín, J.R., Mosquera-Corral, A., Campos, J.L., Méndez, R. (2008). Biofilm and granular systems to improve anammox biomass retention. *Biochem. Eng. J.*, 42, 308–313.

Fernández, I., Dosta, J., Fajardo, C., Campos, J.L., Mosquera-Corral, A., Méndez, R. (2012). Short- and long-term effects of ammonium and nitrite on the anammox process. *J. Environ. Manag.*, 95, S170–S174.

Funamizu, N. and Takakuwa, T. (1994). Simulation of the operating conditions of the municipal wastewater treatment plant at low temperatures using a model that includes the IAWPRC activated sludge model. *Water Sci. Technol.*, 30(4), 105–113.

Gaid, K. (1982). Elimination biologique de l'azote. PhD Thesis, University of Rennes, Rennes.

Ganigué, R., Gabarró, J., López, H., Ruscalleda, M., Balaguer, M.D., Colprim, J. (2010). Combining partial nitritation and heterotrophic denitritation for the treatment of landfill leachate previous to an anammox reactor. *Water Sci. Technol.*, 61, 1949–1955.

Gilbert, J.A., Scott, N.M., Hess, M., Bouskil, N.J., Mason, O.U., Jansson, J.K. (2014). The microbial nitrogen cycling potential is impacted by polyaromatic hydrocarbon pollution of marine sediments. *Aquatic Microbiology*, 5(108), 1–8.

Graham, D.M. and Jolis, D. (2017). Pilot-scale evaluation of pH-based control of single stage deammonification processes for sidestream treatment. *Water Environ. Res.*, 89(2), 99–104.

Hashimoto, S., Furukawa, K., Shioyama, M. (1987). Autotrophic denitrification using elemental sulfur. *J. Fermentation Technol.*, 65(6), 683–692.

Hassan, K. (2018). Dynamique des communautés bactériennes des boues d'épuration lors de leur enrichissement en bactéries Anammox. PhD Thesis, University of Lorraine, Nancy.

Hellinga, C., Schellen, A.A.J.C., Mulder, J.W., Van Loosdrecht, M.C.M., Heijnen, J.J. (1998). The Sharon process: An innovative method for nitrogen removal from ammonium-rich wastewater. *Water Sci. Technol.*, 37, 135–142.

Hellinga, C., Van Loosdrecht, M.C.M., Heijnen, J.J. (1999). Model based design of a novel process for nitrogen removal from concentrated flows. *Math. Comput. Model. Dyn. Syst.*, 5, 351–371.

Hendrickx, T.L.G., Kampman, C., Zeeman, G., Temmink, H., Hu, Z. (2014). High specific activity for anammox bacteria enriched from activated sludge at 10°C. *Bioresour. Technol.*, 163, 214–221.

Henze, M., Grady, C.L., Gujer, W., Marais, G.R. (1986). Activated sludge no. 1. *Water Sci. Technol..*, 29, 1–33.

Henze, M., Harremoes, P., La Cour Jansen, J., Arvin, E. (1995). *Wastewater Treatment – Biological and Chemical Processes*. Springer-Verlag, Berlin.

Hignette, M., Lamort, B., Langouet, M., Leroy, S., Martin, G. (1996). Élimination des nitrates par filtration biologique autotrophe sur soufre en aquariologie marine. In *EUAC Congress*. Institut Océanographique Paul Ricard, 7–13.

Holt, J.G., Krieg, N.R., Sneath, P.H.A, Staley, J.T., Williams, S.T. (1994). *Bergey's Manual of Determinative Bacteriology*, 9th edition. Williams and Wilkins, Baltimore.

Huang, P. and Hogsett, M. (2011). The robustness of anammox communities treating full-scale sidestream municipal anaerobic digester filtrate. *Proc. Water Environ. Fed.*, 2011, 3147–3155.

Jang, A., Minsu, B., Sungyoun, K., Yeonghee, A., In, S.K., Bishop, P. (2005). Assessment of characteristics of biofilm formed on autotrophic denitrification. *J. Microbiol. Biotechnol.*, 15(3), 455–460.

Jaroszynski, L.W., Cicek, N., Sparling, R., Oleszkiewicz, J.A. (2012). Impact of free ammonia on anammox rates (anoxic ammonium oxidation) in a moving bed biofilm reactor. *Chemosphere*, 88, 188–195.

Jetten, M.S., Kuenen, J.G., Strous, M. (1999). Key physiology of anaerobic ammonium oxidation. *Appl Environ Microbiol*, 65(7), 3248–3250.

Jetten, M.S., Wagner, M., Fuerst, J., van Loosdrecht, M., Kuenen, J. G., Strous, M. (2001). Microbiologie et application du procédé d'oxydation anaérobie de l'ammonium ("anammox"). *Curr. Opin. Biotechnol.*, 12(3), 283–288.

Joss, A., Salzgeber, D., Eugster, J., König, R., Rottermann, K., Burger, S., Fabijan, P., Leumann, S., Mohn, J., Siegrist, H. (2009). Full-scale nitrogen removal from digester liquid with partial nitritation and anammox in one SBR. *Environ. Sci. Technol.*, 43(14), 5301–5306.

Jung, J.Y., Kang, S.H., Chung, Y.C., Ahn, D.H. (2007). Factors affecting the activity of anammox bacteria during start up in the continuous culture reactor. *Water Science and Technology: A Journal of the International Association on Water Pollution Research*, 55(1–2), 459–468.

Kabore, A.W.J., Chang, Y., Xie, J., Guo, M. (2022). Research advances in anammox granular sludge: A review. *Crit. Rev. Environ. Sci. Technol.*, 52, 631–674.

Kadam, P.C. and Boone, D.R. (1996). Influence of pH on ammonia accumulation and toxicity in halophilic, methylotrophic methanogens. *Appl. Environ. Microbiol.*, 62(12), 4486–4492.

Kartal, B., Koleva, M., Arsov, R., van der Star, W., Jetten, S.M., Strous, M. (2006). Adaptation of a fresh water anammox population to high salinity wastewater. *J. Biotechnol.*, 126(4), 546–553.

Kartal, B., Rattray. J., van Niftrik, LA., van de Vossenberg., J. (2007). *Candidatus Anammoxo-globus propionicus* a new propionate oxidizing species of anaerobic ammonium oxidizing bacteria. *Syst. Appl. Microbiol.*, 30(1), 39–49.

Kartal, B., Rattray. J., van Niftrik, LA., van de Vossenberg., J., Schmid. M.C., Damsté, J.S., Jetten, M.S.M., Strous, M. (2008). *Candidatus "Brocadia fulgida"*: An autofluorescent anaerobic ammonium oxidizing bacterium. *FEMS Microbiol. Ecol.*, 63(1), 46–55.

Kartal, B., Kuenen, J.G., van Loosdrecht, M.C. (2010). Municipal sewage treatment with anammox. *Science*, 328(5979), 702–703.

Kimura, Y., Isaka, K., Kazama, F., Sumino, T. (2010). Effects of nitrite inhibition on anaerobic ammonium oxidation. *Appl. Microbiol. Biotechnol.*, 86(1), 359–365.

Klaus, S., Baumler, R., Rutherford, B., Thesing, G., Zhao, H., Bott, C. (2017). Startup of a partial nitritation-anammox MBBR and the implementation of pH-based aeration control. *Water Environ. Res.*, 89, 500–508.

Koenig, A. and Liu, L.H. (1996). Autotrophic denitrification of landfill leachate using elemental sulfur. *Water Sci. Technol.*, 34(3–4), 469–476.

Koenig, A. and Liu, L.H. (2001). Kinetic model of autotrophic denitrification in sulfur packed-bed reactor. *Water Res.*, 35(8), 1969–1978.

Koenig, A. and Liu, L.H. (2002). Use of limestone for pH control in autotrophic denitrification: Continuous flow experiments in pilot-scale packed bed reactors. *J. Biotechnol.*, 99(2), 161–171.

Kosgey, K., Zungu, P.V., Bux, F., Kumari, S. (2022). Biological nitrogen removal from low carbon wastewater. *Front. Microbiol.*, 13, 968812.

Kuenen, J.G. (2008). Anammox bacteria: From discovery to application. *Nature*, 6, 320–326.

Kuenen, J.G. and Robertson, L.A. (1992). The use of natural bacterial populations for the treatment of sulphur-containing wastewater. *Biodegradation*, 3, 239–254.

Kuypers, M.M.M., Sliekers, A., Lavik, G., Schmid, M. (2003). Anaerobic ammonium oxidation by anammox bacteria in the Black Sea. *Nature*, 422, 608–611.

Kuypers, M.M.M., Marchant, H., Kartal, B. (2018). The microbial nitrogen-cycling network. *Nat. Rev. Microbiol.*, 16, 263–276.

Lackner, S., Gilbert, E.M., Vlaeminck, S.E., Joss, A., Horn, H., Van Loosdrecht, M.C.M. (2014a). Full-scale partial nitritation/anammox experiences – An application survey. *Water Res.*, 55, 292–303.

Lackner, S., Thoma, K., Gilbert, E.M., Gander, W., Schreff, D., Horn, H. (2014b). Start-up of a full-scale deammonification SBR-treating effluent from digested sludge dewatering. *Water Sci. Technol.*, 71, 553–559.

Ladipo-Obasa, M., Forney, N., Riffat, R., Bott, C., Debarbadillo, C., De Clippeleir, H. (2022). Partial denitrification–anammox (PdNA) application in mainstream IFAS configuration using raw fermentate as carbon source. *Water Environ. Res.*, 94, 10711.

Lam, P., Lavik, G., Jensen, M.M., van de Vossenberg, J., Schmid, M., Woebken, D., Gutierrez, D., Amann, R., Jetten, M.S.M., Kuypers, M.M.M. (2009). Revising the nitrogen cycle in the Peruvian oxygen minimum zone. *Proceedings of the National Academy of Sciences*, 106(12), 4752–4757.

Lavigne, R. (1993a). La dénitrification biologique. *Aquarama*, 130, 27–38.

Lavigne, R. (1993b). La dénitrification biologique : 2e partie. *Aquarama*, 131(3), 42–50.

Le, T., Peng, B., Su, C., Massoudieh, A., Torrents, A., Al-Omari, A. (2019). Impact of carbon source and COD/N on the concurrent operation of partial denitrification and anammox. *Water Environ. Res.*, 91, 185–197.

Lemaire, R., Zeng, R.J., Yuan, Z., Keller, J. (2003). Simultaneous nitrification, denitrification, and phosphorus removal in a lab-scale sequencing batch reactor. *Biotechnol. Bioeng.* 84(2), 170–8.

Lemaire, R., Marcelino, M., Yuan, Z. (2008). Achieving the nitrite pathway using aeration phase length control and step-feed in an SBR removing nutrients from abattoir wastewater. *Biotechnol. Bioeng.*, 100, 1228–1236.

Lemaire, R., Thomson, C., Christensson, M., Zhao, H., Thesing, G. (2013). Mainstream deammonification using Anita™ Mox process W15: Mainstream deammonification and shortcut TN removal – Innovation and implementation. In *Proceedings of WEFTEC 2013*, 6 October, Chicago.

Lesouef, A., Payraudeau, M., Rogalla, F., Kleiber, B. (1992). Optimizing nitrogen removal reactor configurations by on-site calibration of the IAWPRC activated sludge model. *Water Sci. Technol.*, 25(6), 105–123.

Li, M., Ford, T., Li, X., Gu, J.D. (2011). Cytochrome cd1-containing nitrite reductase encoding gene nirS as a new functional biomarker for detection of anaerobic ammonium oxidizing (anammox) bacteria. *Environ. Sci. Technol.*, 45(8), 3547–3553.

Li, X., Klaus, S., Bott, C., He, Z. (2018). Status, challenges, and perspectives of mainstream nitritation–anammox for wastewater treatment. *Water Environ. Res.*, 90, 634–649.

Liao, D., Li, X., Yang, Q., Zhao, Z., Zeng, G. (2007). Enrichment and granulation of anammox biomass started up with methanogenic granular sludge. *World J. Microbiol. Biotechnol.*, 23(7), 1015–1020.

Magrí, A., Matias, B.V., Szögi, A.A. (2012). Anammox sludge immobilized in polyvinyl alcohol (PVA) cryogel carriers. *Bioresour. Technol.*, 114, 231–240.

Martin, G. (1979). *Le problème de l'azote dans les eaux*. Technique et Documentation, Paris.

Martin, K.J. and Nerenberg, R. (2012). The membrane biofilm reactor (MBfR) for water and waste-water treatment: Principles, applications, and recent developments. *Bioresour. Technol.*, 122, 83–94.

Martinelle, K. and Westlund, A. (1996). Ammonium ion transport? A cause of cell death. *Cytotechnology*, 22(1–3), 251–254.

McCartney, D.M. and Oleszkiewicz, J.A. (1990). Carbon and nutrient removal in a sequencing batch reactor at low temperatures. *Environ. Technol.*, 11, 99–112.

Mines, R.O.J. and Sherrard, J.H. (1999). Temperature interactions in the activated sludge process. *J. Environ. Sci. Health*, 34(2), 329–340.

Moon, H.S., Ahn, K.-H., Lee, S., Nam, K., Kim, J.Y. (2004). Use of autotrophic sulfur-oxidizers to remove nitrate from bank filtrate in a permeable reactive barrier system. *Environ. Pollut.*, 129(3), 499–507.

Moraes, B.S., Souza, T.S.O., Foresti, E. (2012). Effect of sulfide concentration on autotrophic denitrification from nitrate and nitrite in vertical fixed-bed reactors. *Process Biochem.*, 47, 1395–1401.

Mosquera-Corral, A., González, F., Campos, J.L., Mendéz, R. (2005). Partial nitrification in a SHARON reactor in the presence of salts and organic carbon compounds. *Process Biochem.*, 40, 3109–3118.

Ni, S.Q. and Meng. J. (2011). Performance and inhibition recovery of anammox reactors seeded with different types of sludge. *Water Sci. Technol.*, 63(4), 710.

Ni, B.-J., Pan, Y., Guo, Y., Virdis, B. (2016). Denitrification processes for wastewater treatment. In *Metalloenzymes in Denitrification: Applications and Environmental Impacts*, Moura, I., Moura, J.J.G., Pauleta, S.R., Maia, L.B. (eds). The Royal Society of Chemistry, London.

Niftrik, L.A., Fuerst, J.A., Sinninghe, J.S., Kuenen, J.G., Strous, M. (2004). The anammoxosome: An intracytoplasmic compartment in anammox bacteria. *FEMS Microbiol. Lett.*, 233(1), 7–13.

Noutsopoulos, C., Mamais, D., Statiris, E., Lerias, A. (2018). Reject water characterization and treatment through shortcut nitrification/denitrification: Assessing the effect of temperature and type of substrate. *J. Chem. Technol. Biotechnol.*, 93(12), 250–261.

Nugroho, R., Takanashi, H., Hirata, M., Hano, T. (2002). Denitrification of industrial wastewater with sulfur and limestone packed column. *Water Sci. Technol.*, 46(11–12), 99–104.

Oh, S.E., Kim, K.S., Choi, H.C., Cho, J., Kim, I.S. (2000). Kinetics and physiological characteristics of autotrophic denitrification by denitrifying sulfur bacteria. *Water Sci. Technol.*, 42(3–4), 59–68.

Oleszkiewicz, J.A. and Berquist, S.A. (1988). Low temperature nitrogen removal in sequencing batch reactors. *Water Res.*, 22(9), 1163–1171.

Oshiki, M., Shimokawa, M., Fujii, N., Satoh, H., Okabe, S. (2011). Physiological characteristics of the anaerobic ammonium-oxidizing bacterium "*Candidatus Brocadia Sinica*". *Microbiology*, 157(6), 1706–1713.

Pan, S.H. (2007). Autotrophic denitrification of groundwater in a granular sulfur-packed upflow reactor. PhD Thesis, University of Texas at Arlington.

Pannard, A., Souchu, P., Chauvin, C., Gross, E. (2017). Définitions scientifiques de l'eutrophisation. In *L'eutrophisation. Manifestations, causes, conséquences, et prédictibilité*, Pinay, G., Gascuel, C., Ménesguen, A. et al. (eds). CNRS, Ifremer, INRA and Irstea, 14–81.

Park, H., Sundar, S., Ma, Y., Chandran, K. (2015). Differentiation in the microbial ecology and activity of suspended and attached bacteria in a nitritation anammox process. *Biotechnol. Bioeng.*, 112, 272–279.

Raboni, M., Torretta, V., Viotti, P., Urbini, G. (2014). Calculating specific denitrification rates in pre-denitrification by assessing the influence of dissolved oxygen, sludge loading and the mixed-liquor recycle. *Environ. Technol.*, 35(20), 2582–2588.

Raboni, M., Gavasci, R., Viotti, P. (2015). Influence of denitrification reactor retention time distribution (RTD) on dissolved oxygen control and nitrogen removal efficiency. *Water Sci. Technol.*, 72(1), 45–51.

Ratel, M.O. and Debrieu, C. (2003). Élimination des nitrates des eaux potables. Technical document, FNDAE, Ministère de l'Agriculture de l'Alimentation, de la Pêche et des Affaires Rurales, Direction de l'Espace Rural et de la Forêt, Paris.

Regmi, P., Miller, M.W., Holgate, B., Bunce, R., Park, H., Chandran, K. (2014). Control of aeration, aerobic SRT and COD input for mainstream nitritation/ denitritation. *Water Res.*, 57, 162–171.

Rysgaard, S., Nøhr Glud, R., Risgaard-Petersen, N., Dalsgaard, T. (2004). Denitrification and anammox activity in arctic marine sediments. *Limnol. Oceanogr.*, 49(5), 1493–1502.

Sahinkaya, E., Kilic, A., Duygulu, B. (2014). Pilot and full-scale applications of sulfur-based autotrophic denitrification process for nitrate removal from activated sludge process effluent. *Water Res.*, 60, 210–217.

Schalk, J. (2000). A study of the metabolic pathway of anaerobic ammonium oxidation. PhD Thesis, University of Delft, Delft.

Schippers, C.J. and Kruithof, C.J. (1987). Removal of nitrate by slow sulfur/limestone filtration. *Aqua*, 5, 274–280.

Schmidt, J.P., Dejoia, A.J., Ferguson, R.B., Taylor, R.K., Young, R.K., Havlin, J.L. (2002). Corn yield response to nitrogen at multiple in-field locations. *Agronomy Journal*, 94(4), 798–806.

Sengupta, S., Ergas, S., Lopez-Luna, E. (2007). Investigation of solid-phase buffers for sulfur-oxidizing autotrophic denitrification. *Water Environ. Res.*, 79, 2519–2526.

Sierra-Alvarez, R., Beristain-Cardoso, R., Salazar, M., Gomez, J., Razo-Flores, E., Field, J.A. (2007). Chemolithotrophic denitrification with elemental sulfur for groundwater treatment. *Water Res.*, 41, 1253–1262.

Sikora, L.J. and Keeney, D.R. (1976). Evaluation of sulphur-*Thiobacillus denitrificans* nitrate removal system. *J. Environ. Quality*, 5(3), 298–303.

Soares, M.I.M. (2002). Denitrification of groundwater with elemental sulfur. *Water Res.*, 36(5), 1392–1395.

Spanjers, H., Vanrolleghem, P., Olsson, G., Dold. H. (1996). Respirometry in control of the activated sludge process. *Water Science and Technology*, 34(3–4), 117–126.

Strous, M., Van Gerven, E., Ping, Z., Kuenen, J.G., Jetten, M.S.M. (1997). Ammonium removal from concentrated waste streams with the anaerobic oxidation (anammox) process in different reactor configuration. *Water Res.*, 31(1), 1955–1962.

Strous, M., Kuenen, J., Jetten, M.S., Key, M. (1999). Physiology of anaerobic ammonium oxidation. *Appl. Environ. Microbiol.*, 65(7), 3248–3250.

Strous, M., Pelletier, E., Mangenot, S., Rattei, T., Lehner, A., Taylor, M.W. (2006). Deciphering the evolution and metabolism of an anammox bacterium from a community genome. *Nature,* 440, 790–794.

Suarez, C., Persson, F., Hermansson, M. (2015). Predation of nitritation–anammox biofilms used for nitrogen removal from wastewater. *FEMS Microbiol. Ecol.*, 91, 124.

Trimmer, M., Nicholls, J.C., Deflandre, B. (2003). Anaerobic ammonium oxidation measured in sediments along the Thames Estuary, United Kingdom. *Appl. Environ. Microbiol.*, 69(11), 6447–6454.

Trouve, C., Chazal, P.M., Gueroux, B., Sauvaitre, N. (1998). Denitrification by new strains of *Thiobacillus denitrificans* under non-standard physicochemical conditions. Effect of temperature, pH and sulfur source. *Environ. Technol.*, 19(6), 601–610.

Van de Graaf, A.A., Mulder, A., De Bruijn, P., Jetten, M.S., Roberstson, L.A., Kuenen, J.G. (1995). Anaerobic oxidation of ammonium is a biologically mediated process. *Appl. Environ. Microbiol.*, 61(4), 1246–1251.

Van Der Star, W.R., Abma, W.R., Blommers, D., Mulder, J.-W., Tokutomi, T., Strous, M. (2007). Startup of reactors for anoxic ammonium oxidation: Experiences from the first full-scale anammox reactor in Rotterdam. *Water Res.*, 41, 4149–4163.

Van Der Star, W.R., Miclea, A.I., Van Dongen, U.G., Muyzer, G., Picioreanu, C., Van Loosdrecht, M.C. (2008). The membrane bioreactor: A novel tool to grow anammox bacteria as free cells. *Biotechnol. Bioeng.*, 101, 286–294.

Van Kempen, R., Mulder, J.W., Uijterlinde C.A., Loodstrecht, M.C.M. (2001). Overview: Full scale experience of the SHARON® process for treatment of rejection water of digested sludge dewatering. *Water Science & Technology*, 44(1), 145–52.

Wang, D., Fu, Q., Xu, Q., Liu, Y., Hao, H., Yang, Q., Zeng, G.., Li, X., Ni, B.-J. (2017). Free nitrous acid-based nitrifying sludge treatment in a two-sludge system enhances nutrient removal from low-carbon wastewater. *Bioresour. Technol.*, 244(1), 920–928.

Water Environment Research Foundation (WERF) (2014). Deammonification. Report, Alexandria.

Wett, B., Podmirseg, S.M., Gomez-Brandon, M., Hell, M., Nyhuis, G., Bott, C. (2015). Expanding DEMON sidestream deammonification technology towards mainstream application. *Water Environ. Res.*, 87, 2084–2089.

Wiemann, M., Schenk, H., Hegemann, W. (1998). Traitement anaérobie des eaux usées de tannerie avec élimination simultanée des sulfures. *Water Res.*, 32, 774–780.

Wiesmann, U. (1994). Biological removal of nitrogen in wastewaters. *Adv. Biochem. Eng./Biotechnol.*, 51, 113–154.

Wiesmann, U., Choi, S.I., Dombrowski, E.-M. (2006). *Fundamentals of Biological Wastewater Treatment*. John Wiley & Sons, Hoboken, NJ.

Wild, S.R., Rudd, T., Neller, A. (1994). Fate and effects of cyanide during wastewater treatment processes. *Sci. Total Environ.*, 156, 93–107.

Winogradsky, M.S. (1890). Recherches sur les organismes de la nitrification. *Annales Institut Pasteur*, 4, 213–231.

Zeng, H. and Zang, T.C. (2005). Evaluation of kinetic parameters of a sulfur-limestone autotrophic denitrification biofilm process. *Water Res.*, 39(20), 4941–4952.

Zhang, T.C. and Lampe, D.G. (1999). Sulfur: Limestone autotrophic denitrification processes for treatment of nitrate-contaminated water: Batch experiments. *Water Res.*, 33(3), 599–608.

Zhang, S.L., Tong, Y.A., Liang, D.L., Lu, D.Q. (2004). Nitrate-N movement soil profile as influenced by rate and timing of nitrogen application. *Acta Pedologica Sinica*, 41, 270–277.

Zhang, L., Narita, Y., Gao, L., Ali, M., Oshiki, M., Okabe, S. (2017). Maximum specific growth rate of anammox bacteria revisited. *Water Res.*, 116, 296–303.

Zhou, W., Sun, Y., Wu, B., Zhang, Y., Huang, M., Miyanaga, T., Zhang, Z. (2011). Autotrophic denitrification for nitrate removal using sulfur-limestone. *J. Env. Sci.*, 23(11), 1761–1769.

2

Phosphorus Removal

Phosphorus (P) emissions from human activity contribute to eutrophication of aquatic ecosystems. The main sources of phosphorus entering rivers are wastewater effluents and agricultural runoff, with 70% of this attributable to wastewater discharge (Boer et al. 2015).

This reality has led to stricter phosphorus discharge standards and increased pressure on the water industry to reduce phosphorus loads entering rivers, particularly in ecologically sensitive areas (European Council 2000; UKTAG 2013). Thus, targeted phosphorus removal has become increasingly common in large urban wastewater treatment plants (STEU). However, sensitive watercourses are sometimes located in remote areas, receiving phosphorus discharges from smaller wastewater treatment plants. Furthermore, wastewater from small communities is often less rigorously treated, and the potential negative impacts of phosphorus release from small treatment plants may be underestimated (Bowes et al. 2015; May et al. 2015).

2.1. The phosphorus cycle

Phosphorus (^{15}P) is a non-metallic chemical element, abundant in nature. It is the 11th constituent of the Earth's crust (0.12%) and belongs to the pnictogen family (15th group of the periodic table) with an oxidation state that varies from -III to +V. In nature, phosphorus combines with other elements to form a large number of minerals. As with nitrogen (N), phosphorus (P) is important for life since it is essential for the synthesis of the nucleic acids RNA and DNA. It is also found in the skeleton of organisms in the form of PO_4. In the primative Earth, all phosphorus was found in igneous rocks. It was through the surface weathering of these rocks on the continents that phosphorus was gradually transferred to the oceans. It has been calculated that it took more than 3 billion years (Ga) to saturate the oceans with the minerals apatite $Ca_5(PO_4)_3F$ and hydroxyapatite $Ca_5(PO_4)_3OH$, which are the primary natural sources of phosphorus. The phosphorus cycle is unique among major biogeochemical cycles: It does not have a gaseous component, at least not in significant quantities, and consequently has virtually no effect on the atmosphere. It also differs from other cycles in that the transfer of phosphorus (P) from one reservoir to another is not controlled by microbial reactions, as is the case, for example, for nitrogen. Phosphorus is a limiting element in several terrestrial ecosystems because there is no large atmospheric reservoir of phosphorus like there is for carbon, oxygen and nitrogen, and its availability is directly linked to the surface weathering of rocks. It is unclear whether this limitation applies to the ocean, but most researchers believe it would over long timescales.

Virtually all terrestrial phosphorus is derived from the weathering of calcium phosphates on surface rocks, primarily apatite. Although soil contains a large volume of phosphorus, only a small portion is accessible to living organisms. This phosphorus is absorbed by plants and transferred to animals through their diet. Some is returned to the soil through animal excrement and dead organic matter. Another portion is transported to the oceans, where a fraction is used by benthic organisms and plankton to secrete their skeletons; the other fraction settles on the ocean floor as dead organisms or particles and becomes part of the sediments. These sediments are gradually transformed into sedimentary rocks through burial; much later, the rocks are brought back to the surface by tectonic movements, and the cycle begins again.

After these rocks are eroded, the phosphorus is carried away to waterways or oceans by runoff (Pinay et al. 2017). In rural areas, the main sources of phosphorus come from fertilizers (organic and inorganic). Human activity interferes with the phosphorus cycle through phosphate mining, largely for fertilizer production. When added to soils in excess, phosphates are leached into aquatic systems. Since phosphorus is often a limiting nutrient in rivers, lakes and coastal marine waters, adding phosphorus to these systems can act as a fertilizer and cause eutrophication problems (high biological productivity resulting from an excess of nutrients). Phosphorus in surface waters (rivers, lakes and oceans) is supplied by both point sources and diffuse sources. The latter account for the majority of phosphorus discharges. Phosphorus loss can be caused by erosion or runoff from fields, with agriculture being the primary source. Lake sediments act as phosphorus reservoirs and also represent a diffuse source. Wastewater from urban areas and industrial discharges (from factories) are point sources of phosphorus pollution in waters. The presence of phosphorus is normal and desirable in the environment, but excessive amounts can be harmful.

2.2. Eutrophication

Soil particles, laden with nutrients, are carried to waterways. Plants then absorb the phosphate in its inorganic form (primarily phosphate ions in the form of $H_2PO_4^-$ and HPO_4^{2-}) for their growth. Humans and other living organisms such as herbivores and carnivores also assimilate phosphate and excrete it in their feces or during decomposition. The phosphate is then reintroduced into waterways and wastewater systems.

When present in excessive quantities in a body of water, phosphorus can have serious consequences for aquatic ecosystems, such as eutrophication. Eutrophication can affect both freshwater (stagnant and flowing) and saltwater (coastal) waters. Eutrophication has been defined (Durrant et al. 1999) as an enrichment of water with nutrients, which results in the excessive proliferation of aquatic plants and algae (algal bloom) to the detriment of certain other species. This leads to a deterioration in water quality. There may be other equally undesirable changes (bad odors, cloudy water, etc.), which impair the use of water for drinking water purposes, recreation (fishing, swimming, water sports, water releases) and industrial activities (transport, electricity generation) (Deronzier and Choubert 2002). The critical concentration of phosphates ($P\text{-}PO4^{3-}$) at which the eutrophication phenomenon begins is approximately 0.1–0.2 mg. L^{-1} for rivers (Huang et al. 2013) and 5–10 µg. L^{-1} in stagnant water (Lemercier et al. 2003). Phosphorus is often identified as the main limiting factor for algal growth. Its role is clearly demonstrated in freshwater aquatic environments where nitrogen is in excess. Conversely, for coastal marine environments, nitrogen (in the form of nitrates) is often identified as the main driver of eutrophication (Nemery 2017).

Phosphorus being a preservative element, combating eutrophication in a watercourse requires addressing as many upstream point sources as possible. The objective is to reach the phosphorus concentration threshold that limits algal growth, depending on the characteristics of the environment.

Excess phosphorus leads to an uncontrolled proliferation of plants or microscopic algae. This occurs in summer when several conditions are met: the presence of nutrients (nitrogen, phosphorus), sunlight, high temperatures and slow water flow. This excess contributes to a deterioration in water quality, disrupting the ecological balance of waterways and their use for both drinking water and recreation.

Given the significant diffuse inputs of nitrogen from agriculture, against which there are very limited means of action, it is not rational to prioritize the reduction of nitrogen compounds to combat eutrophication.

The key factor requiring action is phosphorus. Indeed, during periods of low water levels, when the risk of eutrophication is highest, the main sources of phosphorus discharge are localized and directly linked to human activity (discharges from municipalities and industries). Therefore, by addressing these discharges, it is possible to sufficiently reduce the amount of phosphorus released and hence limit eutrophication. Figure 2.1 provides a simplified summary of the phosphorus cycle. It shows the different sources and modes of transport of this pollutant to the aquifer and surrounding watercourses.

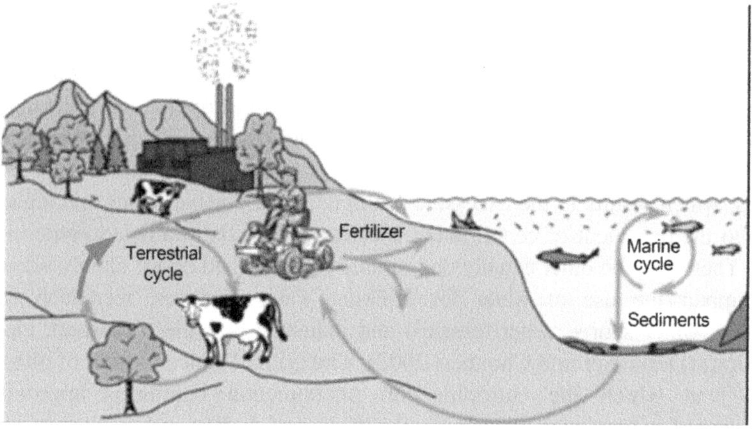

Figure 2.1. *Simplified diagram of the phosphorus cycle in nature*

2.3. Oxidized mineral forms of phosphorus in wastewater

In urban areas, phosphorus comes primarily from municipal wastewater. It is mainly generated by human metabolism, laundry detergents and industrial and agricultural discharge.

In wastewater, approximately 50% of the phosphorus is in the form of soluble inorganic orthophosphates, 35% complex inorganic phosphates (polyphosphates) and 15% organic phosphates present in organic matter. Phosphorus bound in solid particles represents approximately 11–15% of the total phosphorus load.

Inorganic phosphorus is composed of water-insoluble phosphorus such as orthophosphate ions (PO_4^{3-}) – also called the simple inorganic form, whether as free ions or ions associated with organic matter in the form of colloids – as well as the various different anions of phosphoric acid ($H_2PO_4^-$ and HPO_4^{2-}) and polyphosphates. Polyphosphates, also called condensed inorganic phosphorus, make up 45% of total phosphorus. Organic phosphorus, on the other hand, is composed of phospholipids, phosphoproteins, nucleic acids (DNA, RNA), polysaccharide phosphates and soluble phosphorus such as glucose-1-phosphate, adenosine diphosphate, inositol mono- and hexaphosphate, glycerophosphates and phosphocreatine. Organic phosphorus represents only about 10% of total phosphorus (Deronzier and Choubert 2002).

In wastewater, phosphorus exists in two main forms: mineral phosphorus, also called inorganic phosphorus, and organic phosphorus. Together, these two forms represent total phosphorus (Pt). Each form has different reactivities and affinities for chemical precipitation, adsorption and biological dephosphatation processes. Consequently, the effectiveness of phosphorus removal techniques depends considerably on the form in which it is present. However, the most prevalent form of phosphorus is the simple inorganic form due to the biological degradation of the condensed inorganic form in its solid state by microorganisms. Phosphoric acid (H_3PO_4) is a triacid characterized by the acidity constants at 25°C presented below.

Phosphates are the best-known oxides of phosphorus. Phosphates exist in three forms: orthophosphates, polyphosphates ($Na_3(PO_3)_6$, $Na_5P_3O_{10}$, $Na_4P_2O_7$) and organic phosphates.

Phosphoric acid (H_3PO_4) can dissociate into dihydrogen phosphate ions $H_2PO_4^-$, hydrogen phosphate ions HPO_4^{2-} and phosphate ions PO_4^{3-}.

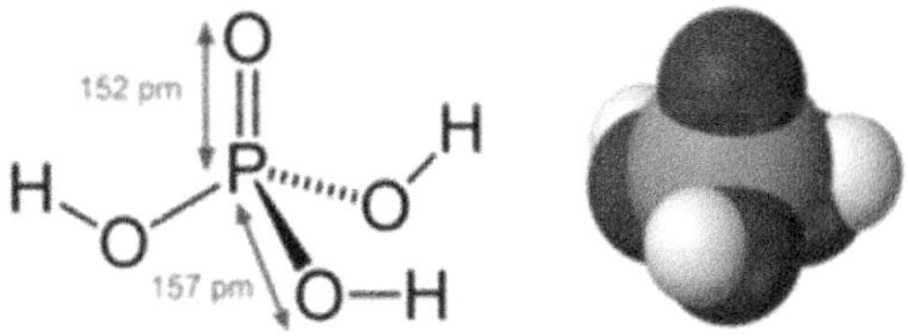

Figure 2.2. *Phosphoric acid*

$H_3PO_4 + H_2O \rightleftharpoons\rightarrow H_2PO_4^- + H_3O^+$	$Ka1 = 7.13 \times 10^{-3}$	$pKa1 = 2.16$
$H_2PO_4^- + H_2O) \rightarrow \rightleftharpoons HPO_{42}^- + H_3O^+)$	$Ka2 = 6.38 \times 10^{-8}$	$pKa2 = 7.21$
$HPO^{42-} + H^2O \rightleftharpoons\rightarrow PO^{43-} + H^3O^+$	$Ka3 = 3.79 \times 10^{-13}$	$pKa3 = 12.32$

These three ions are collectively called orthophosphates. The phosphorus atom exists in its highest oxidation state (+5):

$$P\ total = P\ organic + P\ mineral$$

In urban wastewater, 80% of soluble phosphorus is comprised of approximately 82% of PO_4-P and 18% polyphosphates. Particulate phosphorus (20%) is comprised of approximately 30% organic phosphorus and 70% mineral phosphorus.

Orthophosphates are therefore the salts of orthophosphoric acid H_3PO_4 and ionize depending on the pH in three different chemical configurations.

Therefore, we can distinguish, for example, monosodium phosphates (NaH_2PO_4), disodium phosphates (Na_2HPO_4) and trisodium phosphates (Na_3PO_4). Polyphosphates are molecules containing at least two phosphate units. Their chemical formula is Mn + $2PnO_3n$ + 1.

Phosphorus pentoxide (P_2O_5) is also worth mentioning since agronomists and soil analysis laboratories often express the results of phosphorus measurements in sludge using this oxide. It forms when phosphorus burns in air and reacts very violently with water to produce phosphoric acid. Therefore, understanding the distribution of different orthophosphate and phosphate species as a function of pH is important for determining the presence of solubilization, complex formation and the mechanisms of phosphorus sorption in water (Stumm and Morgan 1996). The diagram shown in Figure 2.3 allows us to determine the speciation of dissolved phosphorus at a given pH. The concentration of the species $H_2PO_4^-$ is elevated for pH levels below 7, whereas the species HPO_4^{2-} is predominant at a pH between 7 and 10. For pH values ranging from 10 to 12, HPO_4^{2-} predominates over PO_4^{3-}. For pH values above 12.5, the concentration of PO_4^{3-} becomes significant and exceeds that of HPO_4^{2-} (Jenkins et al. 1971).

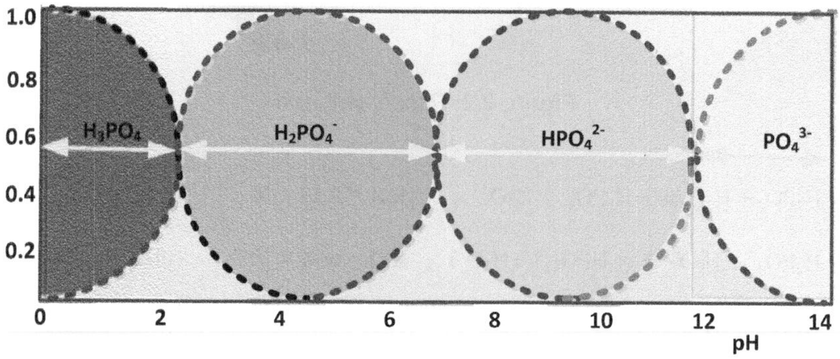

Figure 2.3. *Speciation diagram as a function of pH (Jenkins et al. 1971)*

According to Figure 2.3, the predominant species in wastewater pH (pH ~ 5–8) are HPO_4^{2-} and $H_2PO_4^-$.

Our knowledge of the origin of phosphorus in wastewater stems from knowledge of natural phosphorus sources and its use. The phosphorus comes from:

– Human metabolism: humans excrete between 1 and 2 grams of phosphorus per day. This is the main source of phosphorus in waterways (Dernat et al. 1994).

– Laundry and cleaning products: 1 to 2 grams of P per day per dwelling.

– Industrial discharge: effluents from the food processing industry, slaughterhouses, industrial laundries, surface treatment plants and specialized chemical industries. They carry a quantity of phosphorus compounds roughly equivalent to that of domestic wastewater in relatively urbanized areas.

– Agricultural or naturally occurring runoff is retained in the soil and does not end up in wastewater. Surface waters may contain some due to erosion and runoff carrying soil particles into waterways.

The phosphorus in wastewater, whether particulate or soluble, is essentially composed of:

– inorganic phosphorus (mainly polyphosphates) and orthophosphates, some of which come from the hydrolysis of the former;

– organic phosphorus: phospholipids, esters, polynucleotides, ATP, ADP, phosphorus contained in cell membranes (plant or animal).

Sodium and potassium phosphates are soluble in water; monocalcium phosphates and magnesium phosphates are also soluble, but to a lesser extent. Other phosphates are insoluble. Total phosphorus is the sum of inorganic and organic phosphorus.

Many forms of phosphorus are highly soluble in water and are therefore found in wastewater. Wastewater treatment plants often neglect phosphorus removal. Treated water discharged into the environment is rich in phosphorus and often in nitrates.

Half of the phosphorus is in the form of orthophosphates and water-soluble organic phosphates. Particulate phosphorus is either in organic or mineral form.

The organic fraction corresponds to all the phosphates in animal and plant organic matter (see Figure 2.4), living or undergoing mineralization. The inorganic fraction can be present in two configurations: combined phosphorus (calcium, iron or aluminum salts) and phosphorus fixed or adsorbed to the surface of particles and their constituents (calcium carbonate, iron and aluminum hydroxides, clay, organic matter). Total phosphorus is equivalent to the sum of soluble phosphorus and particulate phosphorus (Nemery 2017).

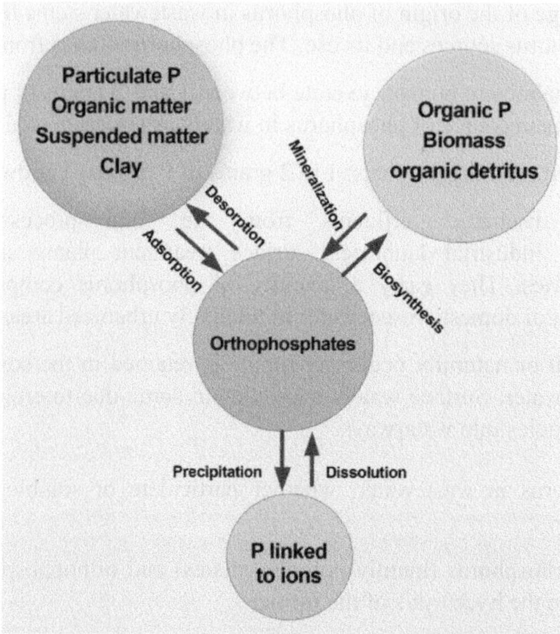

Figure 2.4. *Forms of phosphorus found in wastewater*

Eutrophication of surface waters can be treated primarily by acting on phosphorus, which is often considered a limiting factor.

2.4. Phosphorus removal processes

2.4.1. *Overview*

Assuming that one equivalent inhabitant emits 60 g of BOD_5, the average amount of phosphorus discharged per inhabitant equivalent (survey results) is 3.1 g of P per equivalent inhabitant (PE) in the case of a station supplied by a separate network and 2.6 g of P per PE in the case of a station supplied by a combined network.

Note that these values are significantly lower than the ratio of 4 g of P per PE still frequently used today. This ratio was justified when phosphorus was widely present in detergent composition, whereas today it has been replaced by additives (citrate, silicate or zeolites). Average total phosphorus concentrations vary between 9 and 20 $mg.L^{-1}$. Orthophosphates represent 60 to 85% of the total phosphorus, with the concentration increasing the longer the retention time in the collection network.

Phosphorus removal from wastewater can be achieved using physicochemical methods, biological treatments and/or combinations of both. Numerous large-scale techniques are well established.

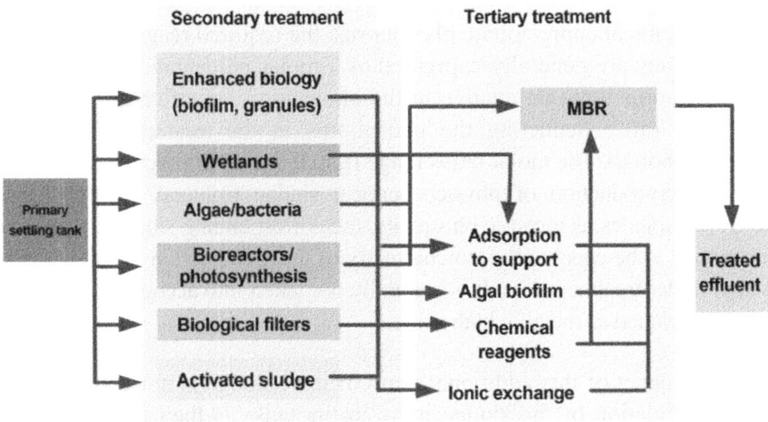

Figure 2.5. *Physicochemical and biological processes for phosphorus removal*

2.4.2. *Biological phosphorus removal*

Biological phosphorus removal is possible with solutions involving free-floating cultures. Activated sludge and wetland biology are solutions that have been successfully tested and implemented at various sites.

An initial assessment reveals the compartmentalization of phosphorus:

– Phosphorus for synthesis: bacterial synthesis also requires phosphorus, which, eliminated simultaneously with the BOD from the water phase, ends up entirely in the sludge produced. The synthetic phosphorus is generally estimated at 1% of the BOD flux eliminated:

P synthesis (Psy) = 1% BOD removed

– Biological phosphorus removal: in this configuration, with the presence of an anaerobic zone to carry out the process, there is no standard calculation method. The required anaerobic volume is generally taken from a range corresponding to a retention time in the zone of 2 to 4 hours. The retention time calculation can be based on the average incoming flow rate, the peak flow rate or the sum of all flow rates, including recirculations, entering the zone. The performance levels used can vary considerably; while a 50% efficiency rate seems to be generally accepted, values of 80% can be found at many sites. Although all the bioremoved phosphorus ends up in the sludge, the conventional approach generally does not take into account the overproduction of sludge by this process.

– Physicochemical coprecipitate phosphorus: the required reagent levels can vary considerably. They are generally expressed as a molar percentage, either relative to the total phosphorus input or relative to the phosphorus specifically to be removed by the process (after deducting the phosphorus in the treated effluent and the synthetic phosphorus). The molar ratios range from 0.5 to 2. Coprecipitated phosphorus leads to an overproduction of physicochemical sludge (mineral sludge). Since the phosphorus precipitates as metallic phosphate, the corresponding sludge overproduction is often assumed to be equal to the stoichiometry of the precipitation reaction, regardless of the stated molar treatment rate. It is generally not taken into account that some of the reactant is precipitated as metallic hydroxide.

Often, the impact of this additional mass of inert sludge is not considered in the sizing. The calculation of the sludge mass in the tanks is then performed without accounting for co-precipitation, and the sludge concentration and volatile fraction are deduced from this. Since the physicochemical removal of phosphorus is calculated afterwards, its impact is only considered in terms of the required reagent quantity and the additional sludge to be treated in the process, but not in terms of the sizing of the activated sludge tanks.

2.4.2.1. *Activated sludge system*

The activated sludge process forms the foundation of biological wastewater treatment, widely recognized for its adaptability in combating the eutrophication of aquatic ecosystems. This method relies on the complex dynamics between microbial communities and operating conditions. Despite its effectiveness, the process is sometimes hampered by performance fluctuations.

Microorganisms that break down carbon and nitrogen incorporate a very small fraction of phosphorus into their cellular material (membrane structure, ATP and RNA). The portion of phosphorus eliminated simply by this pathway represents approximately 25% of the daily amount to be degraded. To increase phosphorus removal performance, activated sludge plants achieve transfer in particulate form through physicochemical precipitation, through biological accumulation within the treatment biomass or through a combination of both processes (combined treatment system).

In both cases, the phosphorus is then removed from the treatment systems via sludge removal.

Polyphosphates are complex compounds that cannot be precipitated. Hence, they are used in detergents. They are broken down in a biological hydrolysis process into orthophosphates in activated sludge. Bacteria convert polyphosphates into orthophosphates, which are then released under anaerobic conditions and reabsorbed under aerobic (and anoxic) conditions.

The biological removal of phosphorus using activated sludge is considered an environmentally sustainable solution. However, experience shows that this solution is subject to fluctuating performance and is highly dependent on skilled operators, leading to difficulties in process control.

The phosphorus requirements for the synthesis of new bacteria are around 1% (i.e., 100/1) of the BOD_5 removed by the biomass present in the biological tank. Assimilation does not allow for extensive phosphorus removal since the BOD_5/P ratio is significantly higher, ranging between 5 and 10. Phosphorus is therefore not an energy source, as the phosphorus compounds in the effluent are not degraded by the activated sludge to provide energy. It can thus be considered a parameter characterizing the formation of cellular material.

In conventional activated sludge (CAS), the organic phosphorus content is around 2% per gram of biomass (expressed as suspended volatile matter), whereas in dephosphatizing activated sludge, the organic phosphorus content can reach 8 to 10% per gram of biomass.

While chemical solutions are based on the precipitation of phosphates with a reagent, biological solutions are based on the principle of phosphorus accumulation in the form of polyphosphates by microorganisms. Biological phosphorus removal is therefore based on the accumulation of phosphorus by microorganisms beyond the "normal" needs of metabolic processes (i.e., absorption), that is, beyond their own metabolic needs for growth.

2.4.2.1.1. Mechanisms involved

Phosphorus (P) accumulates in the form of polyphosphates, which are stored as energy reserves. For this reason, phosphate-accumulating organisms (PAOs) tend to outcompete other organisms. It is now generally accepted that the level of biological P removal is directly proportional to the number of PAOs present in the system.

The exception is the presence of glycogen-accumulating organisms (GAOs), which metabolize volatile fatty acids (VFAs) or other carbon compounds in a manner similar to PAOs, but do not accumulate P (Cydzik-Kwiatkowska and Zielinska 2016).

Regardless of the nature of the mechanisms involved, dephosphatation processes are only achieved when the treatment pathway implements an alternation of anaerobic and aerobic zones.

The phenomena involved are complex. Their implementation requires separating the specific reactions into three compartments: anaerobic, anoxic and aerobic.

Let us recall the definitions of anoxic and anaerobic specific to the field of wastewater treatment:

– Anoxic is characterized by the absence of oxygen and the presence of nitrates.

– Anaerobic is characterized by the absence of oxygen and nitrates.

Anaerobic zone

This is the heart of the reactor. The general mechanism of biological dephosphatation is as follows: under anaerobic conditions, PAO absorbs VFAs and stores them as poly-β-hydroxyalkanoate (PHA), poly-β-hydroxybutyrate (PHB) or poly-β-hydroxyvalerate (PHV) from the easily biodegradable substrate of wastewater and the energy released by the intracellular hydrolysis of polyphosphates (Smolders et al. 1995).

These poly-β-hydroxy compounds are synthesized from VFAs and are carbon compounds that primarily serve as energy reserves. They are polymers comprising monomers with four, five, six and seven carbon atoms.

The energy required for this process is obtained from adenosine triphosphate (ATP), formed from the hydrolysis of internally stored polyphosphate or the breakdown of glycogen. The breakdown of polyphosphate results in the release of orthophosphates into the biological tank water (see Figure 2.6). PHA synthesis under anaerobic conditions requires the equivalent of NADH reduction, which is supplied by the PHA cycle (Comeau et al. 1986; Garcia Martin et al. 2006).

Three phases can be distinguished:

1) A rapid release. This release is not linear with respect to time. The rate of this process is independent of the concentration of readily assimilable carbon such as VFAs. This process requires energy, which is supplied by the breakdown of polyphosphates.

2) The release is slowed down due to the use of carbon substrates that require prior hydrolysis.

3) A slow release, due to cell maintenance. This is secondary or endogenous release. This third form of release is not efficient in that it does not lead to intense reabsorption of phosphorus in the aeration tank.

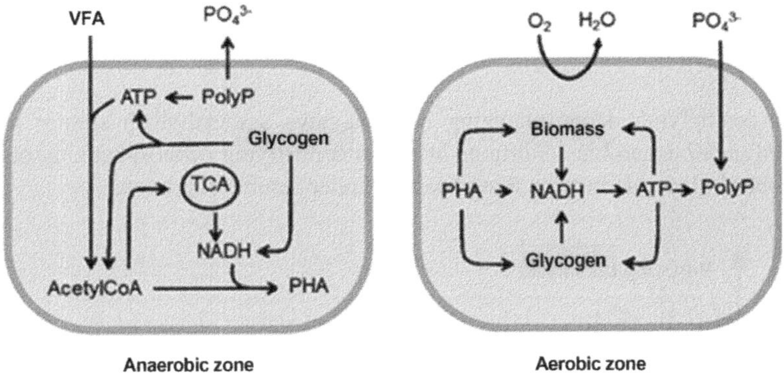

Figure 2.6. *Metabolic pathways of PAO under anaerobic and aerobic conditions. PolyP: polyphosphate; VFA: volatile fatty acid; NADH: nicotinamide adenine dinucleotide; TCA: trichloroacetic acid; ATP: adenosine triphosphate (Smolders et al. 1994)*

Anaerobic conditions are crucial for the absorption of VFAs by PAOs and the release of P. Maintaining sufficient anaerobic retention time and ensuring low nitrate levels is essential, as nitrates can inhibit P release from PAOs.

The consumption of stored PHBs enables the production of energy necessary for bacterial growth, membrane transport of phosphates and their storage in the form of polyphosphates.

Aerobic zone

In the subsequent aerobic stage, these bacteria absorb phosphorus, achieving removal rates of 80 to 90%. All of these mechanisms result in an overaccumulation of intracellular polyphosphates and therefore the removal of phosphates from the liquid phase. In the aeration tank, PHAs and organic matter contained in the wastewater are oxidized by the bacteria. Respiration (using dissolved oxygen) provides the energy needed by bacteria to replenish their polyphosphate stores and grow. Therefore, adequate oxygen levels in the aerobic phase are necessary to support PHA oxidation and P uptake, with a generally recommended dissolved oxygen concentration of approximately 2 mg L^{-1}. During this phase, the accumulated PHA is oxidized by an available electron acceptor, and the resulting energy release is used for P uptake (with stabilizing cations) and glycogen synthesis (Acevedo et al. 2012). The anaerobic zone is therefore considered a selective biological zone for POAs. This zone provides a competitive advantage for POAs, as they can assimilate the substrate in this zone before other microorganisms. In this way, the anaerobic zone offers favorable conditions for the development or selection of a large population of POAs.

Anoxic zone

Non-denitrifying dephosphatizing bacteria have a metabolism similar to that observed under anaerobic conditions, whereas denitrifying dephosphatizing bacteria have a metabolism identical to that observed under aerobic conditions.

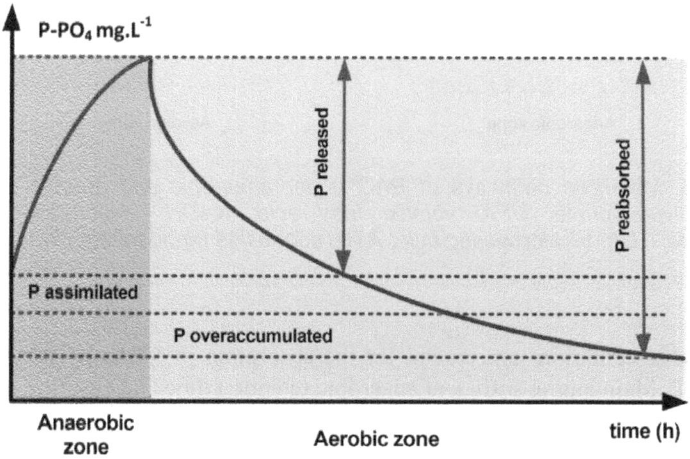

Figure 2.7. *Phosphorus release and reabsorption curve (Deronzier and Choubert 2004)*

Figure 2.7 summarizes the mechanism of biological phosphorus removal. Four types of bacteria coexist within the purifying biomass:

– strict aerobic heterotrophic bacteria, ensuring carbon degradation;

– facultative aerobic heterotrophic bacteria, responsible under anoxic conditions and in the presence of carbon for denitrification (transformation of nitrates into nitrogen gas);

– facultative aerobic heterotrophic bacteria carrying out fermentation, a reaction that transforms organic matter into VFAs under anaerobic conditions;

– autotrophic bacteria with respect to aerobic carbon, responsible for nitrification (transformation of ammonia into nitrates).

Phosphating bacteria are either strictly aerobic bacteria, obtaining energy only from oxygen, or facultatively aerobic bacteria, primarily obtaining energy from oxygen, then from nitrates in the absence of oxygen (Comeau 1997).

Bacteria such as *Candidatus Accumulibacter phosphatis* and *actinobacter* absorb VFAs and release phosphorus into the solution in the absence of oxygen and nitrates. The biochemical mechanisms associated with phosphorus absorption depend on the PAOs and their operating conditions, such as the requirements for carbon, glycogen and electron acceptors, which are constantly adjusted to promote PAO growth and proliferation. *Candidatus Accumulibacter phosphatis* (*Accumulibacter*) is the most studied PAO and forms the basis of most metabolic models. Its taxonomy is now well understood thanks to the acquisition of the complete *Accumulibacter* genome, which has provided information on the expression of genes involved in key metabolic processes.

Alongside *Candidatus Accumulibacter phosphatis*, we find *Actinobacteria* sp. which are also found in abundance in biological systems for removing P. Under anaerobic conditions, they accumulate and store carbon sources such as amino acids and proteins.

2.4.2.1.2. Factors influencing the overaccumulation process

This cyclical process, facilitated by alternating anaerobic and aerobic conditions, is essential for the efficient removal of phosphorus. To maximize the efficiency of this process, specific operating conditions must be optimized. A robust and stable microbial community is essential for good biological phosphorus removal. While the phosphorus content of CAS (expressed in units of dry weight) is around 2%, a phosphorus-removing biomass can contain up to 8% phosphorus.

Other important factors include pH and temperature, with optimal conditions for PAOs including a pH range of 7 to 8 and temperatures between 20 and 30°C, which support the enzymatic activities essential for PAO metabolism. Hydraulic retention time and solid retention time (SRT) are also essential, with an SRT of 10 to 20 days recommended to maintain a stable PAO population and balance the growth rates of PAO and other microbial populations. Strategies such as optimizing anaerobic conditions, managing carbon/P ratios and adjusting the SRT can help suppress GAO activity, which competes with PAO but does not contribute to P removal (Xue et al. 2014).

The effectiveness of biological phosphorus removal is primarily linked to:

– the concentration of simple, easily biodegradable substrates contained in wastewater or that may be generated anaerobically;

– the BOD/P (or COD/P) ratio of wastewater, which affects assimilation efficiency;

– the existence of actual anaerobic and aerobic conditions in the reactors concerned;

– ensuring sufficient retention times in the anaerobic and aerobic reactors for the release and reabsorption reactions to be complete;

– the sludge age.

The main parameters for controlling biological phosphorus depletion are optimizing the functioning of the anaerobic zone and the influence of soluble COD.

Organic carbon from wastewater

The carbon source usable by phosphorus-depleting bacteria is readily biodegradable COD. It is composed primarily of VFAs (molecules with a maximum of six carbon atoms) such as acetic acid, butyric acid and propionic acid. The volatile acids present in wastewater generally result from the fermentation of larger molecules.

The production of these compounds is linked to the very nature of the carbon substrate of the effluent to be treated. The propensity for release is proportional to the amount of readily biodegradable substrate present in the soluble fraction of the untreated wastewater. Consequently, the fraction of COD and/or BOD initially present in the untreated wastewater determines the sizing of the anaerobic zone and the treatment process efficiency. If the easily biodegradable soluble fraction of carbon pollution is low, the retention time in the anaerobic zone must be increased to allow the hydrolysis of other forms of carbon pollution into "simple" carbon substrates necessary to initiate the release process.

Glycogen has been identified as a key player in biological phosphate degradation mechanisms. It plays a role at four levels:

– as a carbon source (as well as VFAs) for the synthesis of PHA;

– as an energy source, in addition to the energy released by polyphosphate hydrolysis;

– as a source of reducing power, meaning that it plays a role in the formation of the specialized molecule nicotinamide adenine dinucleotide (NADH), the presence of which is essential;

– in the synthesis of PHA.

Nitrates

In anaerobic conditions, nitrates disrupt or even inhibit the release process, just as dissolved oxygen does. Acetogenic bacteria use nitrates as a final electron acceptor, inhibiting the fermentative pathways that produce acetate. On the other hand, readily assimilable carbon is consumed primarily by denitrification reactions and then is no longer available for the release mechanism. Several hypotheses have been formulated.

The NO_3^- ion blocks certain metabolic pathways of release:

– The presence of nitrates promotes the growth of denitrifying bacteria at the expense of dephosphate-forming bacteria.

– The bacteria do not experience anaerobic stress because oxygen is bound to nitrates. Therefore, there is no activation of polyphosphate kinase.

– The permissible nitrate content in the anaerobic zone depends on the carbonaceous pollution input and the retention time. The reduction of the anaerobic zone due to the establishment of the anoxic zone, created by nitrate recycling, can be compensated either by a higher COD, which allows for faster denitrification and increases stress, or by a longer retention time. The permissible nitrate values in the anaerobic zone as a function of the inlet COD are (Davelaar 1993):

- 4.5 mg.L^{-1} for a COD of 300 mg.L^{-1};
- 2 mg.L^{-1} for a COD of 160 mg.L^{-1};
- 1 mg.L^{-1} for a COD of 140 mg.L^{-1}.

2.4.2.1.3. The limits of dephosphatation

The effectiveness of biological phosphorus removal depends on several parameters that regulate its function and control its performance.

P/BOD$_5$ ratio

Based on experience, biological phosphorus removal allows the elimination of 3.5 to 4.5 mg of phosphorus per 100 mg of BOD5 consumed, depending on the characteristics of the water being treated. Limiting this to 3.5% results in only 50 to 60% phosphorus removal for typical urban wastewater using biological methods. Therefore, to meet more stringent standards, mixed solutions should be considered where the portion of phosphorus not biologically removed is co-precipitated by the addition of reagents. Figure 2.8 shows, as a function of the P/BOD$_5$ ratio of the untreated water, the percentage of phosphorus that can be removed by biological overaccumulation (based on ΔP/ΔBOD equal to 3.5% and 4.5%) and the degree of co-precipitation that will be necessary to achieve the desired low standards (Vachon et al. 1992).

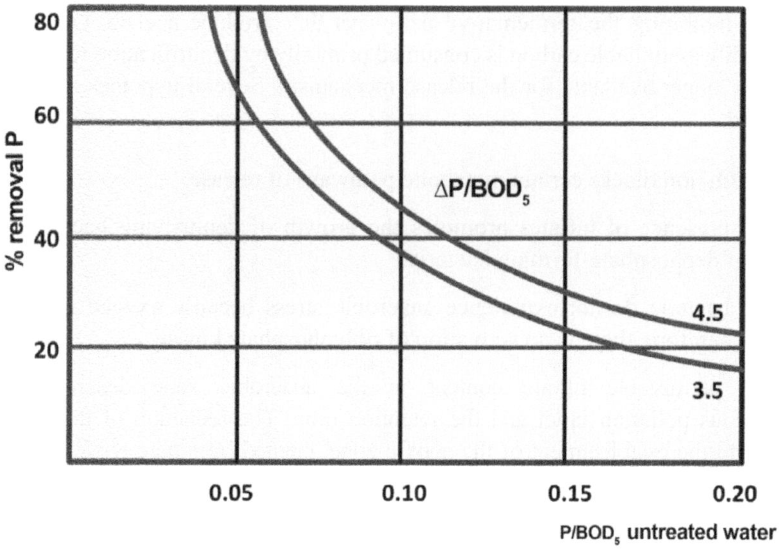

Figure 2.8. *Phosphorus removal efficiency according to the P/BOD5 ratio of wastewater*

Concentration of soluble COD in wastewater

Proper phosphate degradation depends on a sufficient supply of carbonaceous pollution to ensure strict anaerobic conditions and maximize phosphate release in the anaerobic zone. The release kinetics are expressed in mg of P-PO_4 per g of MVS and per day in the anaerobic zone, which is between 40 and 60 when dephosphatation conditions are good.

These values correspond to a soluble COD greater than 400 mg.L^{-1} and to a total COD value greater than 800 mg.L^{-1} for the wastewater. To ensure effective phosphorus removal, wastewater must contain a minimum quantity of VFA between 25 and 100 mg.L^{-1} for each mg of P removed.

Often, the P/COD ratio is used as a reference. The lower this ratio, the more phosphorus the sludge will absorb. A ratio of 129 (mg.mg^{-1}) gives an efficiency of 73%, whereas a ratio of 33.6 only results in a 15% efficiency.

In the case of a wastewater treatment plant connected to a combined sewer system, water quality varies over time. During periods of rain, the concentration of VFAs in the wastewater decreases, as does the phosphorus removal efficiency. There is no reliable criterion that allows for real-time and indirect determination of the amount of VFA in the wastewater available for biological phosphorus removal.

In the case of large-capacity wastewater treatment plants, VFA production can be promoted by installing anaerobic fermenters or primary sedimentation/thickening tanks at the inlet of the plant.

Retention time

This parameter varies under anaerobic conditions depending on supply conditions and the dissolved oxygen concentration. For high biodegradable COD, a relatively short retention time in the anaerobic compartment is sufficient. It can be one hour if the COD is high and two hours for average COD concentrations.

pH

pH is an important factor governing biological phosphorus depletion and plays a crucial role during the anaerobic phase. A low pH leads to a decrease in the ratio of phosphorus released to absorbed acetate. Indeed, the increased pH gradient across the bacterial cell membrane requires additional energy for acetate transport. Part of the energy released by polyphosphate hydrolysis is therefore dedicated to this task, whereas under optimal pH conditions, it would be used for PHA storage. A pH of 6.8 ± 0.7 is an acceptable range (Liu et al. 1996).

Temperature

An increase in temperature promotes the biological removal of phosphorus. Three reasons explain this phenomenon:

– Increased rates of phosphorus release/reabsorption, substrate consumption and bacterial growth. Phosphorus absorption and release mechanisms may be slowed at low temperatures (8–10°C).

– Substantial increase in the quantity of rapidly biodegradable substrates (VFA) with temperature.

– Improved efficiency of nitrification/denitrification processes, with a positive impact on dephosphatation.

Phosphate release under anaerobic conditions

The amount of phosphorus absorbed is directly related to the amount of phosphorus released, and thc greater the overconsumption of phosphorus, the greater the release (Vachon et al. 1992). A very good correlation is obtained between the release of P and the overall assimilated quantity (see Figure 2.9). The deduced relationship gives:

$$\text{P assimilated} = 1.05 \text{ P released} + 90$$

Ammonification of nitrogen in anaerobic conditions

Simultaneously with good phosphorus release, favorable anaerobic conditions promote nitrogen ammonification. A strong correlation between the two phenomena is observed, and it is the optimization of the anaerobic zone that allows for the intensification of both phosphorus release and nitrogen ammonification.

Sludge recirculation rate

Excessive sludge recirculation flow rates should be avoided as they contribute to diluting incoming carbon pollution and reducing retention time in the anaerobic zone.

Sludge age

The shorter the sludge age, the higher the sludge production relative to the amount of pollution removed. This also contributes to better phosphorus removal in the sludge. To achieve optimal phosphorus removal conditions, the sludge age chosen must be strictly necessary for nitrification, and this choice must take into account temperature, which influences nitrification kinetics.

To simplify, we can choose, for example, a winter setting and a summer setting, each with a certain safety margin. For sludge older than 35 days, biological phosphorus removal is no longer effective (Ekama et al. 1983). The phosphorus - accumulation capacity of biomass is then saturated, and P release increases due to endogenous activity (Wentzel et al. 1986). When the sludge is younger, the reaction rate is faster. A sludge age of approximately 20 days is compatible with good biological phosphorus removal efficiency.

Control of nitrification–denitrification

This is an important prerequisite for biological phosphorus removal. Incomplete denitrification leads to the recirculation of nitrates into the anaerobic zone via the returning sludge. Nitrates have a well-known inhibitory effect on phosphorus release, making precise control of denitrification in biological phosphorus removal absolutely essential.

Concentration of dissolved oxygen in the aeration tank

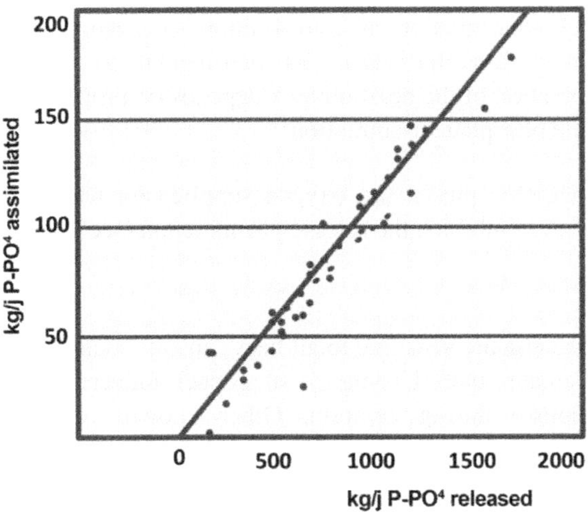

Figure 2.9. *Relationship between the quantity of phosphate assimilation and removal (Vachon et al. 1992)*

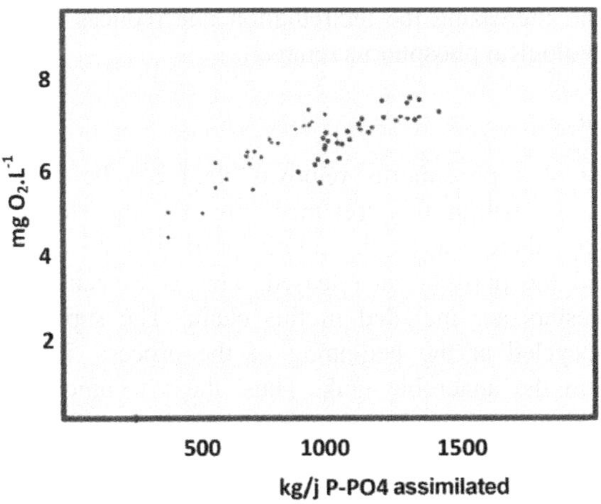

Figure 2.10. *Phosphates assimilated according to the quantity of oxygen supplied (Vachon et al. 1992)*

The oxygen supply in the aeration tank must not be limiting for the reabsorption of phosphates released in the anaerobic zone to occur properly. A sudden drop in oxygen supply leads to a sharp decrease in phosphorus assimilation. Similarly, a sudden increase in oxygen also very quickly results in a marked increase in phosphorus degradation. An O_2 concentration of 2 to 4 $mg.L^{-1}$ is recommended. However, a concentration lower than 0.4 $mg.L^{-1}$ is detrimental to phosphorus removal. Optimizing the operation of the aeration tank depends on proper aeration conditions, which will facilitate phosphate assimilation.

During the anaerobic phase, the oxygen supplied by wastewater or returning sludge reduces, or even inhibits, the release phenomenon (see Figure 2.10).

Metals

Phosphating bacteria are sensitive to all heavy metals such as Cr, Zn, Hg and Ni. A concentration greater than 1.5 $mg.L^{-1}$ of copper induces a definite inhibition of phosphorus uptake during aeration. Other known toxicity limits towards phosphorus-depleting biomass are 5 $mg.L^{-1}$ for cadmium and 5 $mg.L^{-1}$ for nickel and zinc (Villebrun 1989). However, the addition of key precipitation reagents such as ferric chloride does not appear to inhibit the biological removal of phosphorus.

Recirculation of clarifier sludge

It is essential that the sludge does not remain in the secondary settling tank for too long and that it is recirculated quickly to prevent the release of phosphates into the treated effluent. Increasing the recirculation rate reduces release and increases the efficiency of biological phosphorus removal.

Sludge treatment

The efficiency of a phosphorus removal plant can be linked to the sludge treatment process. If, within this treatment process, the sludge is subjected to anaerobic conditions, there is a risk that a large portion of the stored phosphates will be hydrolyzed and the phosphorus released. This is especially true if anaerobic thickeners or digesters are included in this chain. The supernatants from these systems, being recycled at the beginning of the process, impose an additional phosphorus load in the anaerobic tank. Thus, these technical recommendations, when properly implemented, lead to significant treatment efficiencies in biological phosphorus removal. The activated sludge process is a robust solution for the biological removal of phosphorus, often achieving removal efficiencies exceeding 80%. Its effectiveness is influenced by various factors, including bacterial populations, the availability of organic matter and operational parameters. Through monitoring, advanced control and the potential for polyphosphate recovery and recycling, this solution presents a sustainable approach to wastewater management.

In summary, strict anaerobic conditions must be ensured in the phosphorus release zone; non-limiting aeration conditions must be maintained in the aerated zone, allowing for good phosphorus reabsorption and good control of nitrification–denitrification; the design and operation of the sludge treatment chain should avoid any phosphorus recycling; and control of clarification should be achieved through the use of low hydraulic loads and rapid sludge recovery. The composition of the wastewater to be treated is also a crucial parameter, as it determines both the performance and reliability of the biological phosphorus removal process. Concentrated wastewater leads to high phosphorus removal rates (>70%), while less concentrated wastewater requires additional technical intervention through the injection of chemical reagents.

The solution of combining biological and chemical techniques seems to be gaining ground at many sites where biological ones alone are insufficient. It can allow for the combination of the respective advantages of the two techniques: reliability and high yields for one, and more economic use of reagents and low sludge production for the other.

2.4.2.2. *Sizing of biological dephosphatation*

2.4.2.2.1. Anaerobic tank

Optimizing the time spent in the anaerobic tank is essential for the proper functioning of the biological reaction.

The volume of the anaerobic tank (V_{AN}) is written as:

$$V_{AN} = t\,(Qe + Qr)$$

with:

– Qe: average water flow rate during dry weather ($m^3.h^{-1}$);

– Qr: recirculation flow rate ($m^3.h^{-1}$);

– t: retention time (h).

For a recirculation rate of 100%, we obtain:

$$V_{AN} = 2t.Qe$$

In cases where the wastewater treatment plant is supplied by a separate sewer system and the wastewater is normally concentrated, a contact time of two hours in the anaerobic wastewater tank is sufficient.

The tank volume is therefore written as:

$$V_{AN} = 4 * \text{hourly volume during dry weather}$$

Under unfavorable conditions (e.g. very dilute raw water, low COD/P ratio, uniform network draining clear parasitic water), the retention time is extended to 3 or even 4 h.

Thus, at the average dry-weather flow rate and for a recirculation rate of 100%, we obtain:

V_{AN} = 6 * average hourly volume during dry weather

2.4.2.2.2. Anaerobic tank for biological phosphorus removal

This tank must ensure true anaerobic conditions; namely, an $N.NO_3$ concentration upstream of the anaerobic tank less than 0.3 mg. L^{-1}. It is therefore necessarily preceded by an anoxic zone.

Criteria:

– Retention time = 2.4 to 4 hours, calculated from the influent flow rate (without recirculation).

– Sludge age > 4 days.

– $V_{anaerobic} > 10\%\ Q\ (m^3.d^{-1})$ with Q = nominal daily flow rate for a period equal to the sludge age.

– In the absence of precise data on the frequency of rainfall events, one can consider: $V_{anaerobic} > 15\%\ Q\ (m^3.d^{-1})$ with Q = nominal daily flow rate.

The V_{anoxic} calculated is the total volume of the anoxic tanks, the volume of each being proportional to the flow of $N.NO_3$ allowed in each zone.

If the $N.NO_3$ concentration in the influent is negligible, one has:

$V_{anoxic\ 1}/V_{anoxic\ 2} = Q_{recycled\ sludge}/Q_{mixed\ liquid\ recycling}$

When $N.NO_{3influent}$ increases or when $(COD/N)_{influent}$ is low, $V_{anoxic\ 1}$ needs to be increased compared to $V_{anoxic\ 2}$.

It is necessary to verify that the volume of the anaerobic and anoxic tanks does not exceed 40% of the total volume of the tanks.

2.5. Treatment processes incorporating biological dephosphatation

There are many technologies used by manufacturers to remove phosphorus biologically.

2.5.1. *Activated sludge*

Numerous processes are used in activated sludge treatment. Configurations vary depending on the layout of the anaerobic tank, the methods of internal and external sludge recirculation, the number of anoxic/aerobic tanks and the staged or alternating supply system.

2.5.1.1. *Modified Bardenpho process*

The Bardenpho process has been implemented to remove nitrogen. However, in the initial process, a significant release of phosphates was observed in the final tank. This was the first evidence of the detrimental role nitrates have on the release of orthophosphates. In order to create the bacterial stress necessary for dephosphatation, the South African Barnard (1976) added an anaerobic zone upstream of the treatment process with sludge recirculation from the clarifier. This was called the Phoredox process.

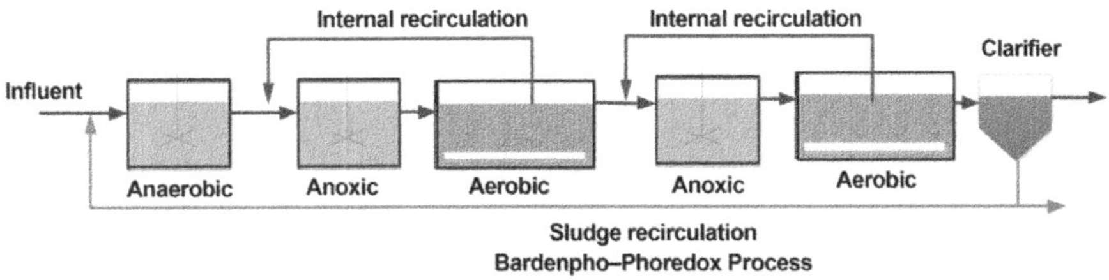

Figure 2.11. *Bardenpho–Phoredox process*

2.5.1.2. *Modified Phoredox process I*

Simpkins noted that the denitrification rate in the second anoxic stage was much lower than in the first denitrification stage. The second stage was eliminated, and the volume of the first denitrification stage was increased (Simpkins et al. 1978).

2.5.1.3. *Modified Phoredox process II*

Then, Barnard (1985) proposed a new modification to the Phoredox process by recirculating the settled sludge to an initial anoxic compartment rather than to the anaerobic one. A portion of the aerated biomass is directed to the second anoxic tank. In order to stimulate bacterial stress and release phosphates, anoxic sludge from the first tank is directed to the anaerobic tank. The COD/biomass ratio is high in the anaerobic tank due to the supply of untreated wastewater. Since the nitrate content supplied is lower, there is no longer competition and biological phosphorus removal can be carried out.

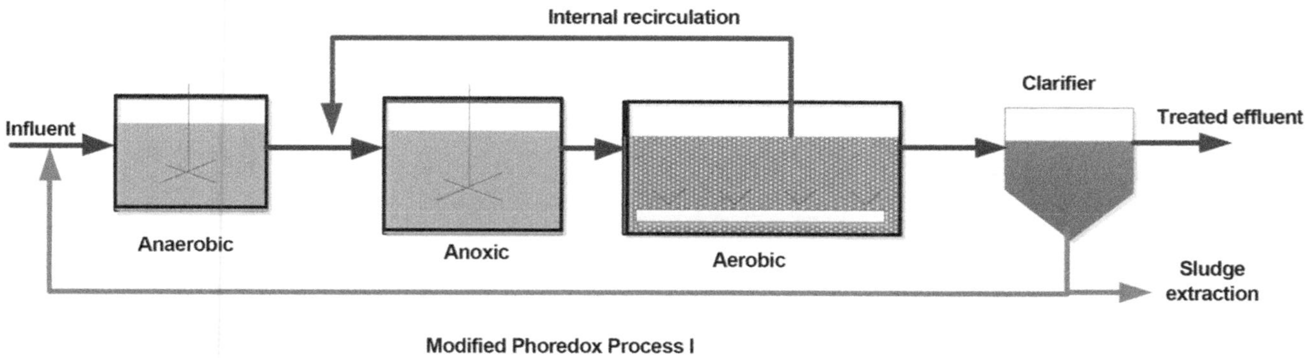

Figure 2.12. *Schematic diagram of the modified Phoredox process I*

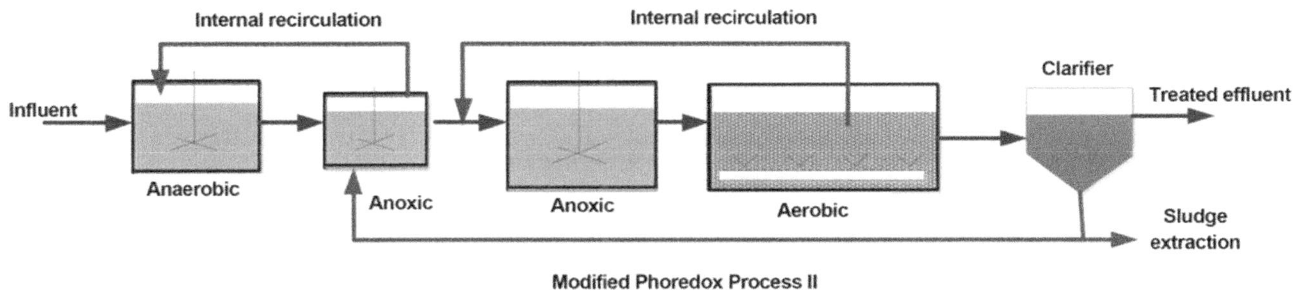

Figure 2.13. *Schematic diagram of the modified Phoredox process II*

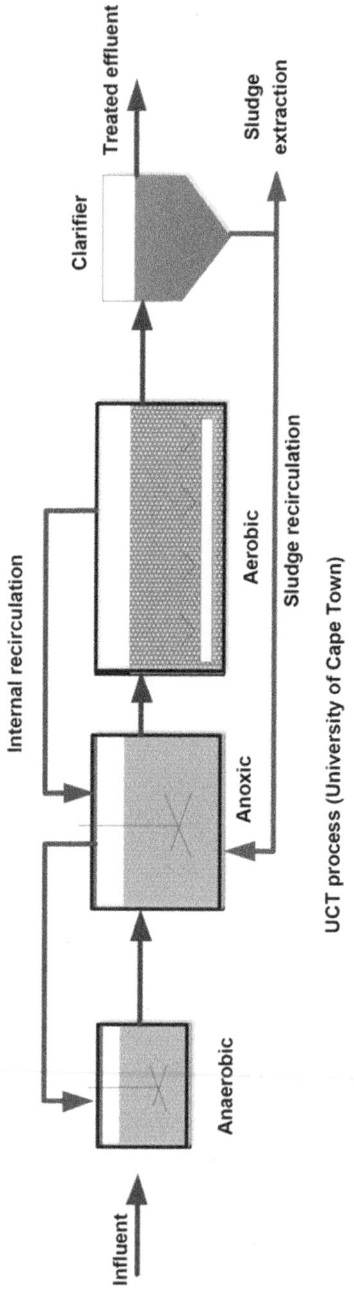

Figure 2.14. *UCT process*

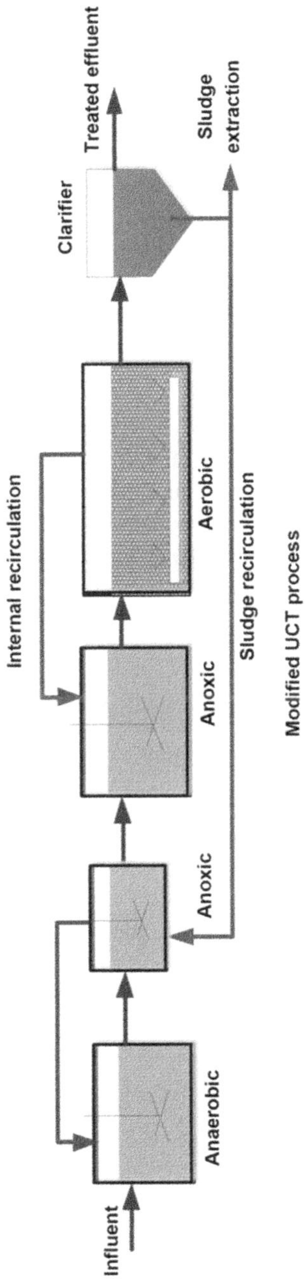

Figure 2.15. *Modified UCT process*

2.5.1.4. *UCT process*

When the NTK/COD ratio is high, denitrification efficiency is insufficient and denitrification is incomplete. Researchers at the University of Cape Town proposed the modified UCT process in which the settled sludge is recirculated to the anoxic tank rather than the anaerobic one (Van Haandel 1981).

Three recirculation loops are integrated: settled sludge to the anoxic tank, biomass from the aerated tank to the anoxic tank and anoxic biomass to the anaerobic tank. Recirculation rates are adapted for each compartment. The nitrate concentration in the recycled sludge becomes significant and sufficient to load the anoxic tank to the maximum quantity it can remove under constant load and flow conditions.

2.5.1.5. *Modified UCT process*

When the TKN COD ratio is very high, the three-loop recirculation principle is maintained. However, an anaerobic tank and a new anoxic tank are placed at the beginning of the treatment process. The new anoxic tank now receives only settled sludge, and the recirculation it directs to the anaerobic tank allows for phosphorus removal without the presence of nitrates.

2.5.1.6. *A/O and A^2/O processes (Veolia)*

These are two modular, multi-compartment processes.

2.5.1.6.1. Anoxic/oxygenated (A/O) process

This solution does not treat phosphorus, but it has the advantage of being modular; that is, the activated sludge tank is divided into compartments that can be aerated or not, giving it operational flexibility.

The A/O process inspired manufacturers to modify and adapt it for phosphorus removal.

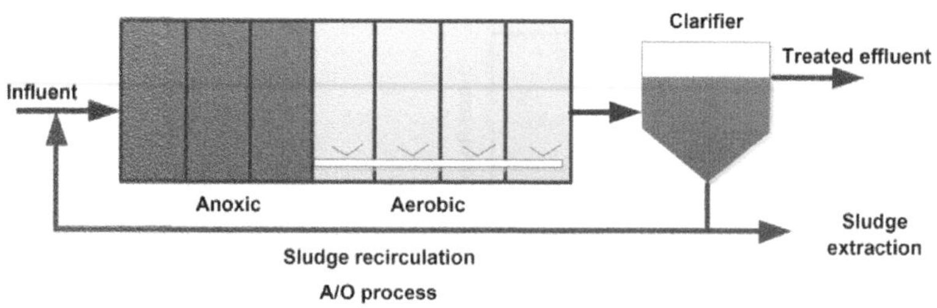

Figure 2.16. *Schematic diagram of the A/O process*

2.5.1.6.2. Anoxic²/oxygenated (A²/O) process

A^2/O activated sludge systems consist of three compartments: anaerobic, anoxic and aerobic for nitrification, denitrification and dephosphatation. This process features internal sludge recirculation from the aerated tank to the anoxic tank. This allows the denitrification process to operate optimally. The recirculation of settled sludge and raw wastewater supplies the anaerobic tank located upstream of the treatment process. Phosphorus is released into the anaerobic tank using the organic matter from the wastewater. The integration of the anaerobic zone gives it new properties and allows it to incorporate phosphorus removal. This process is very similar to the modified Phoredox process.

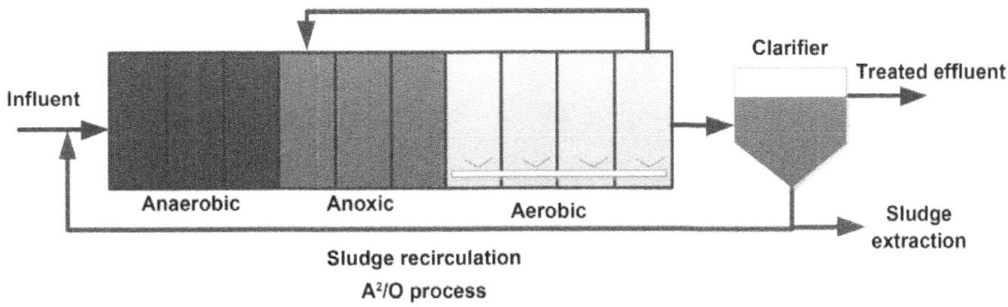

Figure 2.17. *A^2/O process*

2.5.1.7. *Biodenipho™ process*

Developed by Veolia, this process is derived from the Biodenitro™ process regarding nitrogen removal (see section 1.5); to achieve biological phosphorus removal, an anaerobic selector is installed upstream of the oxidation channels and sequentially supplies the biological tanks. The sludge recirculated from the clarifier is directed directly to the anaerobic compartment.

The distributor located at the outlet of the anaerobic tank (1) is a motorized valve that changes position from right to left at a predetermined frequency, towards zone two. Slow-speed submerged mixers keep the bacterial flocs in suspension during the anoxic phase of the process. The motorized weir (3) changes position from high to low, or vice versa, depending on the time allotted for denitrification. The denitrified water flows into the second channel (4) where the degradation of organic carbon and nitrification take place. Surface aerators (6) ensure the oxygenation conditions required for biochemical processes. The dissolved O_2 concentration is monitored using probes (5) distributed throughout the channel. The treated water overflow (7) is lowered alternately with the first overflow (3) to allow the treated water to be discharged.

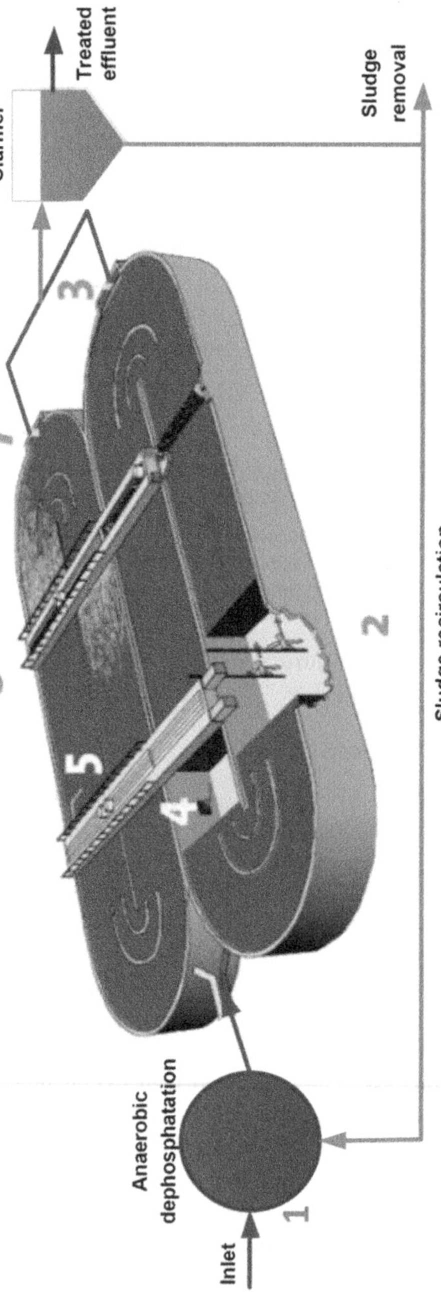

Figure 2.18. *Biodenipho™ process*

2.5.2. *Biofilters: introduction of reagents into the primary settling tank*

Phosphorus removal in an anaerobic biofilter is quite complex due to clogging and the need for frequent backwashing. Therefore, phosphorus is removed through a supplementary treatment.

Since the primary settling tank is essential upstream of the biofilters to reduce the suspended solid load and prevent clogging, the introduction of a chemical reagent significantly reduces suspended solids and also precipitates phosphorus. Care must be taken to ensure sufficient phosphorus remains in the downstream treatment structures to meet the nutritional needs of the bacteria.

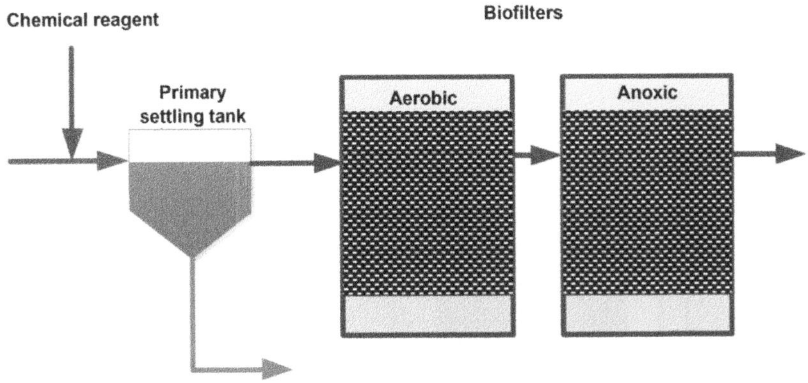

Figure 2.19. *Chemical removal of phosphorus through primary settling*

A flotation device, a rotary sieve or possibly a clarifier can be used in refining with the injection of a chemical reagent. The clarifier is particularly recommended when, in addition to the chemical reagent, activated carbon is added to remove micropollutants.

2.5.3. *Other biological phosphorus removal technologies*

Other technologies are available for the biological removal of phosphorus. These include the incorporation of an anaerobic zone in membrane bioreactors (MBRs), granular sludge reactors and sequential biofilm reactors (SBRs). All are effective if the operating conditions are met.

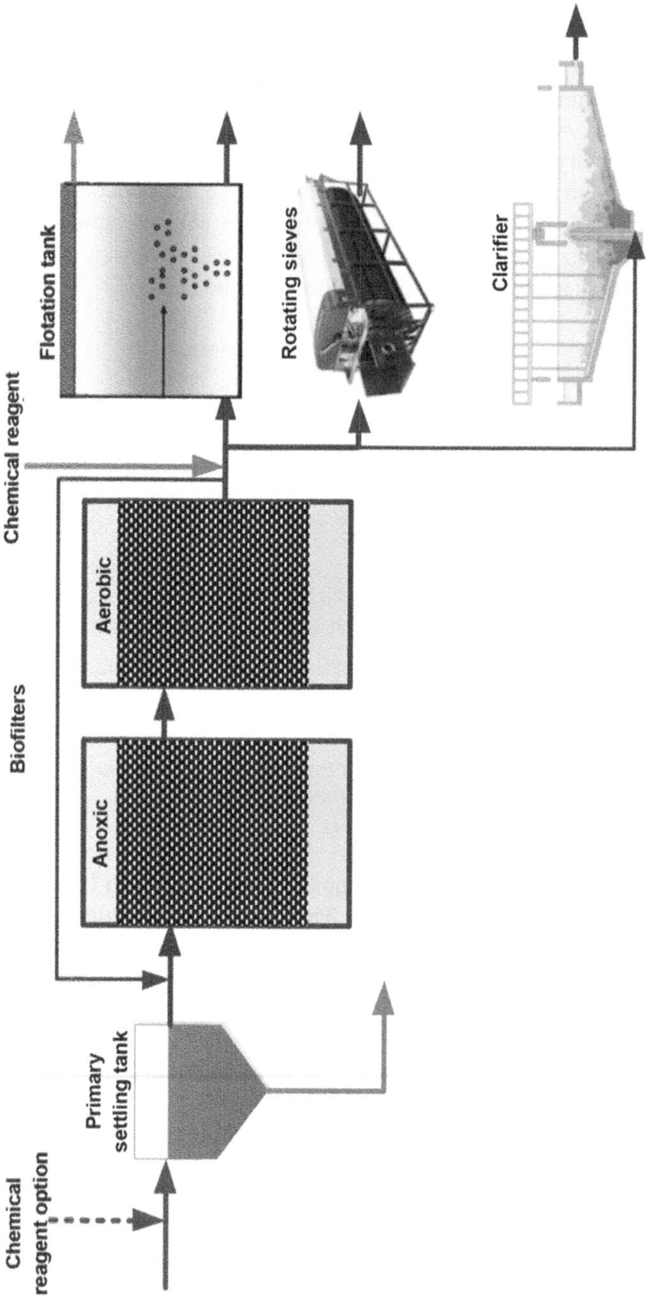

Figure 2.20. *Chemical removal of phosphorus in a refinary*

Granular sludge reactors have been developed to operate under aerobic or anaerobic conditions. Aerobic granular sludge is a suspended biofilm technology that relies on biomass growth on granules rather than flocculation, as in activated sludge plants. A commercial version of this technology is the Nereda system, which has achieved phosphorus removal rates of up to 87% (Pronk et al. 2015).

2.5.3.1. *Membrane bioreactor (MBR)*

MBRs offer several advantages, including the retention of solids within the reactor, resulting in high concentrations of suspended solids in the mixed liquor (Ng et al. 2000). They represent a sophisticated approach to wastewater treatment, combining biological degradation processes with membrane filtration. In MBR systems, biomass degradation occurs in a biological tank, followed by biomass separation using a membrane.

MBR technology effectively eliminates organic and inorganic pollutants, making it suitable for municipal and industrial applications.

Phosphorus removal in MBR systems is possible because these systems retain PAOs in the bioreactor due to their size and compatibility with microfiltration membranes. Several innovative approaches have been developed to improve the efficiency of phosphate removal in MBRs:

– Batch sequencing moving-bed membrane bioreactor: this approach has demonstrated significant improvements in carbon, nitrogen and phosphorus removal, with an average TP removal efficiency of 84.1% under specific operating conditions (Yang et al. 2010).

– Algae-based MBRs: these systems exploit algae-induced phosphate precipitation for the chemical removal of P, producing a P-rich algal biomass. Algae-facilitated chemical removal of P has also been studied with high-density *Chlorella emersonii* cultures within MBRs. A 66 ± 9% removal of total P is achieved from algal biomass containing significant amounts of extracellular and intracellular P, suggesting that algal-induced phosphate precipitation is essential for P removal (Xu et al. 2014).

– Anoxic/oxic membrane bioreactor: capable of simultaneously performing nitrification and denitrification in addition to phosphorus removal, this system highlights the role of PAOs in achieving high phosphorus removal efficiencies (with average removal efficiencies of 94.6% for COD and 90% for PO_4^{3-} P). The presence of PAO in the reactor contributes to the high efficiency of phosphorus removal (Fu et al. 2009).

Membrane bioreactors offer a promising solution for the efficient removal of phosphorus from wastewater, supported by various biological processes and innovative designs. By controlling environmental conditions, membrane bioreactors ensure not only high-quality effluents but also open up avenues for resource recovery, making them a viable option for modern wastewater treatment strategies.

2.5.3.2. *Artificial wetlands*

Artificial wetland systems were introduced in the late 1960s and have since proliferated worldwide. Wastewater treatment now uses a wide variety of artificial wetlands. Although intermittently fed vertical flow systems have rapidly developed and expanded due to the increasing need for ammonia removal, subsurface flow systems are often constructed with horizontal flow. Consequently, the filter bed can be better oxygenated, improving nitrification performance.

P removal in wetlands is achieved by mechanisms such as sorption, precipitation and absorption by plants, with varying yields between 40 and 60% (Vymazal 2007). The substrate used plays a crucial role in the removal process. For example, laterite, a material rich in iron and aluminum, has shown a phosphorus removal potential of up to 99% in laboratory tests and pilot wetland trials. Similarly, natural pyrite used as a substrate can improve the long-term removal of phosphorus and nitrogen dioxide without hindering plant growth or other removal processes, achieving average removal rates of 87.7% for phosphorus and 69.4% for nitrogen dioxide (Wood and McAtamney 1996; Ge et al. 2019).

Artificial wetlands represent a versatile and sustainable approach to wastewater treatment, offering not only high phosphorus removal efficiency but also the potential for integration into diverse operational environments. Their continued development and implementation underscore the importance of constructed wetlands in achieving global water treatment goals.

2.5.3.3. *Microalgae/fungi*

Phosphorus (P) is an essential nutrient for algal growth, and under certain conditions, algae will absorb more P than they require for growth. In such circumstances, the P is absorbed as orthophosphate and stored as polyphosphate granules for use as a growth reserve. In the event of phosphorus deficiency in the environment, when inorganic orthophosphate is unavailable, algae will absorb organic phosphorus, converting it to orthophosphate at the cell surface via the enzyme phosphatase (Larsdotter 2006). The green microalgae *Scenedesmus* sp. and *Chlorella* sp. are known for their ability to significantly absorb phosphorus from the natural environment, and it has been reported that they effectively remove phosphorus from wastewater in various contexts (Gonzalez et al. 1997; De-Bashan et al. 2002; Shi et al. 2007; He and Xue 2010). For example, high levels (up to 90%) of phosphorus removal have been achieved by immobilizing these microalgae on a synthetic substrate, either in sheets or as beads. The cultivation and application of microalgae in wastewater treatment has attracted considerable attention. Microalgae offer the dual advantage of generating biomass that removes nutrients from wastewater, thus contributing to the mitigation of eutrophication in natural water bodies. Incorporating microalgae into wastewater treatment can significantly improve the efficiency of N and P removal from wastewater, from 39 to 66% and 32 to 89%, respectively (Tang et al. 2018). Despite these advantages, the cost of biomass harvesting, which can represent up to 30% of total production expenses due to the small size (5 to 30 μm) and colloidal stability of microalgae cells, presents a significant challenge. An innovative solution involves using flocculation with filamentous fungi such as *Irpex lacteus*, *Trametes versicolor* and *Pleurotus ostreatus* and soil fungi such as *Trichoderma reesei*, which have demonstrated their potential for degrading pharmaceuticals in wastewater and exhibit high enzymatic activity. This approach suggests a promising avenue for harvesting microalgae from wastewater and producing lignocellulose-degrading enzymes for bioethanol production.

Microalgae biofilms have shown remarkable capabilities in removing P, achieving target P concentrations in effluents from 0.15 $mg.L^{-1}$, proving effective in nutrient management. Co-immobilization techniques combining microalgae with bacteria that promote microalgae growth have further improved nutrient removal from municipal wastewater. For example, the co-immobilization of species of *Chlorella* with *Azospirillum brasilense* has considerably improved P removal rates (up to 36% P removal in 6 days) compared to microalgae alone. Furthermore, microalgae biofilms in novel photobioreactors achieved up to 97% (over 24 h) of total P removal under continuous artificial lighting.

Research into granular microalgae-bacteria sludge processes has also been promising, with one study showing that approximately 86% of the influential phosphorus could be eliminated in 6 hours, including a 2-hour dark phase and a 4-hour light phase.

Efficient removal of phosphorus is facilitated by specific genera of microalgae such as *Pantanalinema*, which accumulate poly-P in their cells. Furthermore, research has identified certain strains of microalgae, such as *Botryococcus braunii*, which were capable of removing more than 75% of N and P from treated domestic wastewater while accumulating significant amounts of lipids. The impact of light wavelengths on microalgae growth and nutrient removal was also investigated, revealing that specific light mixing ratios can significantly improve biomass production rates by 45% and P removal efficiency by 90% (Boelee et al. 2011; Patel et al. 2012; Kim et al. 2013; Ji et al. 2020).

2.6. Selectors

Several factors can either promote or inhibit the growth of microorganisms that make up the mixed liquor and influence sludge settling. In fact, it is possible to selectively direct the presence of flocculating bacteria rather than filamentous bacteria by creating conditions conducive to their development. To do this, the recirculated sludge is brought into contact with the tributary substrate in a restricted space called a selector. The selector can be aerobic, anoxic or anaerobic.

In addition to improving sludge settling, separators can also remove total nitrogen (by denitrification) and phosphorus, depending on whether they are anoxic or anaerobic. For domestic wastewater, the total separator volume generally ensures a contact time of around 20 to 60 minutes based on the average tributary flow rate. The separator is often subdivided into three compartments. The first two are generally the same size and each corresponds to a quarter of the total separator volume. The recirculated sludge is returned to the first compartment.

Figure 2.21 shows an activated sludge process with a three-compartment selector.

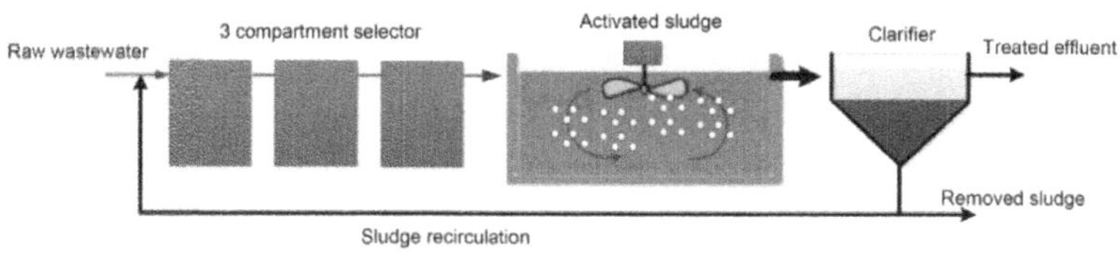

Figure 2.21. *Three-compartment selector*

Table 2.1 presents the recommended BOD/X ratios for the design of different types of selectors.

Compartment	Aerobic selector $gCOD.g^{-1}VSS.d^{-1}$	Anoxic selector $gCOD.g^{-1}VSS.d^{-1}$	Anaerobic selector $gCOD.g^{-1}VSS.d^{-1}$
1	10–12	6	12
2*	6	3	6
Total	3–6	1.5	3

* Includes the volume of the first compartment.

Table 2.1. *Recommended COD/VSS ratio for different types of selectors (Metcalf & Eddy 2014)*

2.6.1. *Aerobic selector*

When the selector is aerobic, the dissolved oxygen concentration should be maintained at 1 or 2 $mg.L^{-1}$ (WEF and ASCE/EWRI 2018). The amount of oxygen required typically corresponds to 15–20% of the removed soluble COD content. Aeration should be designed to meet a specific oxygen utilization rate of at least 65–80 mg $O_2.g^{-1}$ $VSS.h^{-1}$ (MOE 2008). The oxygen requirements of the activated sludge process are not reduced by the presence of this type of selector.

2.6.2. *Anoxic selector*

The anoxic selector is used upstream of an activated sludge process that nitrifies, so that the recirculation of the mixed liquor allows for denitrification. This type of selector offers the following advantages:

– removal of total nitrogen (pre-denitrification);

– reduction in oxygen demand (credit of approximately 2.86 g $O_2.g^{-1}$ NO_3 nitrified);

– increase in alkalinity required for nitrification in activated sludge (credit of approximately 3.5 g $CaCO_3.g^{-1}$ N-NO_3 denitrified). Nitrates must be maintained at a sufficient concentration in the selector to ensure anoxic conditions and denitrification.

For domestic wastewater, the specific nitrate utilization rate is typically 5 to 10 mg N-$NO_3.g^{-1}$ $SS.h^{-1}$ at 20°C (Jenkins et al. 2004). The mixing in the selector should be performed mechanically. If it is carried out by aeration, it is necessary to ensure that the concentration of dissolved oxygen remains below 0.5 mg $O_2.L^{-1}$. It is still preferable for this concentration to be as low as possible.

When the anoxic selector has only one compartment, the BOD/SS ratio recommended for the design should be $\leq$ 1.0 g $BOD_5.g^{-1}$ SS. d^{-1} if T $\leq$ 18°C and $\leq$ 1.5 g $BOD_5.g^{-1}$ SS. d^{-1} if T >18°C.

Another option for sizing a single-compartment anoxic selector is to calculate the average retention time of the biomass in anoxic mode using the following equation. This retention time should be between 1 and 2 days:

ts (d) = SS anoxic (kg SS)/SS removed (kg $SS.d^{-1}$)

The reported sludge volume index (SVI) is approximately 60 to 100 $ml.g^{-1}$ for three-compartment anoxic selectors and 100 to 120 $ml.g^{-1}$ with a single selector (Metcalf & Eddy 2014).

2.6.3. *Anaerobic selector*

The anaerobic selector allows for the biological removal of phosphorus. The retention time is around 0.75 to 2 hours based on the average flow rate to promote the formation of acetate necessary for dephosphatation (WEF and ASCE/EWRI 2018). The compartments are equipped with a mechanical stirring device to keep the mixed liquor in suspension without the introduction of oxygen. No nitrates or dissolved oxygen should be present in the selector (MOE 2008). It is therefore important to properly design and operate an anoxic or anaerobic selector to limit the introduction of dissolved oxygen (WEF and ASCE/EWRI 2018). Furthermore, there are certain filamentous bacteria whose development is favored by the operating conditions of the selectors. In the presence of such organisms, Jenkins et al. (2004) propose several solutions such as the implantation of an anaerobic zone, an anoxic zone and an aerobic zone in series; maintaining a minimum retention time for the treatment requirements; or finally maintaining a concentration of more than 2 mg $O_2.L^{-1}$ in all aerobic zones of the process (and particularly at the inlet of the bioreactor where the effluent from the anoxic or anaerobic selector arrives).

In the presence of a clarifier at the end of the treatment process, the sludge concentration in the aeration tank must be at least 2,000 $mg.L^{-1}$ in terms of VSS. Recirculation rates must be able to be modified by 25% up to 150% or 200% relative to the wastewater flow rate at the inlet in order to maintain the suspended solid concentration in the biological reactor.

The main criteria for sizing and operating a bacteria selector are presented below.

Contact area:

– Sludge age: $G > 4$ days.

– Floc load: $C_F > 100$ mg COD/g TSS.

– Retention time: $t_S = 15$ to 30 minutes.

The value of the floc load C_F allows us to determine the fraction of recirculated sludge from the clarifier Q_1 which must be introduced into the contact zone:

$$CF = \frac{QEB}{Q1}\frac{C\,COD}{CR} \quad \text{and} \quad Q_{BR} = Q_1 + Q_2$$

with:

– C COD: COD concentration of untreated water ($mg.L^{-1}$);

– CR: concentration of recirculated sludge (MES in $g.L^{-1}$).

The retention time is that of the influencing material plus the fraction of recirculating sludge introduced. It allows the tank volume to be calculated:

$$V = t_S \text{ x } (Q_{BR} + Q_1)/60$$

EXAMPLE.

– Data:

C_F = 100 mg DCO/g MESTQ_{EB} = 750 $m^3.h^{-1}$

C_{DCO} (untreated water) = 600 $mg.L^{-1}$

Q_{BR} = 750 $m^3.h^{-1}$

C_{MEST} (recirculated sludge) = 10 $g.L^{-1}$

t_S = 20 minutes

– Calculation:

$$Q1 = \frac{QEB}{CF} \text{x} \frac{CDCO}{CMES} = \frac{750}{100} \text{x} \frac{600}{10} = 450 \text{ m}^3\text{h}^{-1}$$

$$Q_2 = Q_{BR} - Q_1 = 750 - 450 = 300 \text{ m}^3.\text{h}^{-1}$$

$$V = 20 \text{ x } (750+450)/60 = 400 \text{ m}^3$$

Be careful not to confuse the contact zone with the ventilated tank, also called the contact tank, in a contact-stabilization process.

2.7. Physicochemical treatment

Physicochemical processes for phosphorus removal have been used for many years to control phosphorus levels. Precipitation using chemical reagents is a process that allows for the simple removal of phosphorus from wastewater.

The principle involves transforming the phosphorus in the water from a dissolved form to a solid form, which is then separated from the water by sedimentation, flotation or filtration.

These processes are generally reliable and efficient, but they are not without limitations. For example, some can affect the pH of the effluent, many require the addition of chemicals before final discharge, and they often require additional treatment steps due to the production of foreign solids during processing.

Most processes include precipitation, sorption and/or ion exchange mechanisms and are based on chemical dosing to promote phosphorus precipitation.

Phosphorus removal can be achieved by adding salts:

– in untreated wastewater;

– in CAS reactors;

– in the secondary clarifier or at its outlet.

Alternatively, further refinement with a flotation device or rotary sieves is also possible when the suspended solids concentration is low.

The physicochemical treatment consists of transforming orthophosphates (dissolved) into particulate form (floc) by forming a low-solubility precipitate through the addition of metallic salts. The resulting precipitate is then separated from the effluent by separation processes, either by sedimentation (the most common solution) or by specific equipment (flotation, filtration, rotary sieves).

There are various metallic salts capable of precipitating dissolved phosphorus: aluminum sulfate, ferric chloride, calcium hydroxide.

With the addition of metal ions in primary, tertiary or simultaneous treatment in biological tanks (pre-, post- or co-precipitation), excess ions injected into the tanks form oxides. These oxides remain in the sludge and are therefore mostly recirculated in the tanks. In the event of a phosphorus pollution spike, if ion injection becomes insufficient, the oxides will revert to ions and become available for physicochemical treatment.

Since phosphorus precipitates as metallic phosphate, the corresponding sludge overproduction is often assumed to be equal to the stoichiometry of the precipitation reaction, regardless of the stated molar treatment rate. It is generally not taken into account that some of the reactant precipitates as metallic hydroxide.

Often, the impact of this additional mass of inert sludge is not considered in the sizing. The calculation of the sludge mass in the tanks is then performed without considering co-precipitation, and the sludge concentration and volatile fraction are deduced from this. Since the calculation of the physicochemical removal of phosphorus is performed afterwards, its impact is only taken into account in terms of the quantity of reagent required and the additional sludge to be treated in the sludge line, but not in the sizing of the activated sludge tanks.

2.7.1. *Fundamental principles of phosphorus precipitation*

Since this element is found primarily in the form of mineral anions, these are likely to form insoluble compounds with divalent or trivalent cations. In practice, the ionic species used are Fe^{2+}, Fe^{3+}, Al^{3+} and Ca^{2+} with the aim of forming insoluble precipitates of hydroxo-metallic phosphates.

Three mechanisms can come into play depending on the form of P and the pH:

– precipitation of metallic phosphates from orthophosphates;

– adsorption of organic phosphates;

– coagulation of wastewater colloids that can trap phosphorus.

2.7.1.1. *Metallic coagulants*

The most commonly used metallic coagulants fall into two categories: those based on aluminum and those based on iron.

The popularity of the two coagulants (aluminum sulfate and ferric chloride) is based as much on their effectiveness (under optimal operating conditions) as on their attractive cost.

However, these two coagulants have some limitations, such as:

– pH adjustment before and after treatment, combined with pre-remineralization for soft water;

– ferric chloride is sensitive to cold water;

– higher sludge production for ferric chloride;

– concentrations are sometimes high depending on the chemical composition of the water to be treated.

When dissolved, aluminum sulfate and ferric chloride dissociate very rapidly to form hydrated reaction products. Thus, the metal ion forms the following compounds: $\{Al(H_2O)_6\}^{3+}$ and $\{Fe(H_2O)_6\}^{3+}$.

2.7.1.1.1. Aluminum-based coagulants

Aluminum sulfate is often used in wastewater treatment. The flocs formed by aluminum hydroxide are less dense and settle less well than those formed by iron salts but have the advantage of being colorless. Commercial aluminum salts are generally characterized by their aluminum content, expressed in % Al_2O_3 (representative of the active "matter" contained), and by the "basicity" of the product, expressed by the molar ratio $(OH^-)/3(Al^{3+})$, which corresponds to the fraction of sites occupied by hydroxyl groups relative to the total number of available sites.

The basicity is the degree of pre-neutralization of hydrogen ions during hydrolysis. It is calculated by the following equation, where [OH]/[M] is the molar ratio of the hydroxide bound to the metal ion and ZM is the charge of the metal species:

$$\text{basicity } (\%) = \frac{100 \text{ x } [\text{OH}]}{[\text{M}]\text{ZM}}$$

Basicity affects the ratio between polynuclear and mononuclear species present in the solution and also provides an indication of the coagulant's alkali consumption.

The aluminum ion used for phosphorus precipitation is combined with sulfate, sodium, chloride or hydroxide ions in reagents such as aluminum sulfate ($Al_2(SO_4)_3$), sodium aluminate ($Na_2OAl_2O_3$), aluminum polychlorides (Al(OH)xCly), aluminum polymers and mixed aluminum and iron polyhydrochlorides. While these products are used differently in drinking water, only aluminum sulfate is the most widely used in wastewater among all aluminum-based coagulants.

Aluminum sulfate [$Al_2(SO_4)_3$, nH_2O]; n = 14–18

This is probably the most widely used and one of the most economical. Aluminum sulfate is produced by reacting caustic soda with bauxite or another aluminum-rich mineral (e.g. kaolin), followed by acidification with sulfuric acid.

Its Al_2O_3 content typically varies between 8.5 and 17%. It is corrosive and available in solid or liquid form. Aluminum sulfate dissociates and reacts with water to form a range of soluble mononuclear and polynuclear species (containing one or more aluminum ions), as well as an amorphous aluminum hydroxide precipitate (Van Benschoten and Edzwald 1990).

The basic reaction, when adding an aluminum salt to water, is the precipitation of aluminum hydroxide and the release of acid:

$$Al_2(SO_4)_3 \rightarrow 3\ SO_4^{2-} + 2\ Al^{3+}$$

$$2Al^{3+} + 6\ H_2O \rightarrow 2Al(OH)_3 + 6\ H^+$$

The solubility product of $Al(OH)_3$ is: $[Al^{3+}]\ [OH^-]^3 = 10^{-33}$.

The precipitation of $Al(OH)_3$ leads to the trapping of particles suspended in water, with sizes around one micron.

The mode of action of aluminum salts is based on the fact that the Al^{3+} ion is hydrated by six water molecules, forming an octahedral ligand. Its efficacy depends on the degree of aluminum hydrolysis. The hydrolysis reaction begins immediately after the aluminum sulfate has been dosed into the water. As with any chemical reaction, the degree of hydrolysis depends on the raw materials, reaction time, reaction temperature, and so forth.

Depending on the pH, hydrolysis reactions occur successively. For example:

$$Al(H_2O)_6{}^{3+} + H_2O \rightarrow [Al(H_2O)_5OH]^{2+} + H_3O^+$$

$$[Al(H_2O)_5OH]^{2+} + H_2O \rightarrow [Al\ (H_2O)_4(OH)_2]^+ + H_3O^+$$

$$[Al(H_2O)_4(OH)_2]^+ + H_2O \rightarrow Al(H_2O)_3(OH)_3 + H_3O^+$$

$$Al(H_2O)_3(OH)_3 + H_2O \rightarrow [Al(H_2O)_2(OH)_4]^- + H_3O^+$$

These successive reactions form hydroxylated complexes and cause the aluminum to precipitate as a relatively stable floc. In equilibrium with solid $Al(OH)_3$, the dissolved species are essentially Al^{3+}, $Al(OH)^{2+}$, $Al_2(OH)_2{}^{4+}$ and $Al(OH)_4{}^-$.

Furthermore, aluminum salts can polymerize by forming oxo bridges and create species such as $Al_6(OH)_{15}{}^{3+}$, $Al_7(OH)_{17}$, $Al_8(OH)_{20}{}^{4+}$ or $Al_{13}(OH)_{34}{}^{5+}$ depending on the surrounding conditions. The isoelectric point of the system is around pH 6.

The theoretical effective hydrolysis pH range for alum is from 5.5 to 7.7, with optimal pH values close to 6.0. At the optimal pH, solubility is minimal and the maximum amount of coagulant is converted into solid floc particles. As the pH decreases below 6, positively charged, dissolved aluminum species are formed. When the pH is above 6, the concentrations of negatively charged aluminum species increase (Pernitsky and Edzwald 2006).

The optimal pH range in a particular case depends on the physicochemical composition of the wastewater. Aluminum hydroxide precipitates are positively charged when the pH is below 8 and negatively charged otherwise. Similarly, soluble hydrolysis products are positively charged when the pH is acidic. The average positive charge of soluble alum reaction products with water increases with decreasing pH.

If the pH is too high (above pH 8), aluminum is complexed in the form $Al(OH)_4^-$. For noticeably colored soft water, the optimal pH of $Al_2(SO_4)_3$ is most often between 6 and 6.5, but for cloudy water with moderate alkalinity, it will probably be between 6 and 7.

Because the quality of untreated water varies in terms of suspended solids, alkalinity and organic matter, the pH range for coagulation is maintained between 6 and 6.5 to avoid high concentrations of residual aluminum in the treated water. This is particularly noticeable when the alkalinity and/or pH of the untreated water are high. Very high $Al_2(SO_4)_3$ concentrations are then necessary to reach the optimal coagulation pH.

The alternative is to add a controlled amount of acid to lower the pH to a lower value before adding a measured amount of coagulant. However, alkalis (lime, sodium carbonate or caustic soda) are titrated to increase the pH of slightly alkaline waters. The hydrogen ions will react with the water's alkalinity and, in doing so, decrease the water's pH, as shown in the equation for aluminum sulfate:

$$Al_2(SO_4)_3{,}18H_2O \rightarrow 2Al^{3+} + 3SO_4^{2-} + 18H_2O \rightarrow 2Al(OH)_3 + 6H^+ + 3SO_4^{2-} + 12H_2O$$

Simplified solubility diagrams (see Figure 2.22) were developed to describe the relationships between metal salt concentrations and pH. They can help describe the different species of metal salts present in water at varying pH levels. This information is useful when choosing the coagulant type and concentration.

This is explained by the curves surrounding the amorphous precipitate zone of aluminum hydroxide, which indicate the equilibrium concentrations of the soluble species in equilibrium with $Al(OH)_3(s)$.

For a given concentration of coagulant (expressed in Al^{3+}), the curves delineate the pH range where the products of aluminum sulfate hydrolysis are most readily available (as precipitates). Outside this range, these products are soluble, disrupting the coagulation mechanism.

A change in temperature causes a change in the position of the curves surrounding the amorphous precipitate zone of aluminum hydroxide. As the temperature decreases, the pH for the minimum solubility of aluminum increases, as shown in Figure 2.22 (Crittenden et al. 2005).

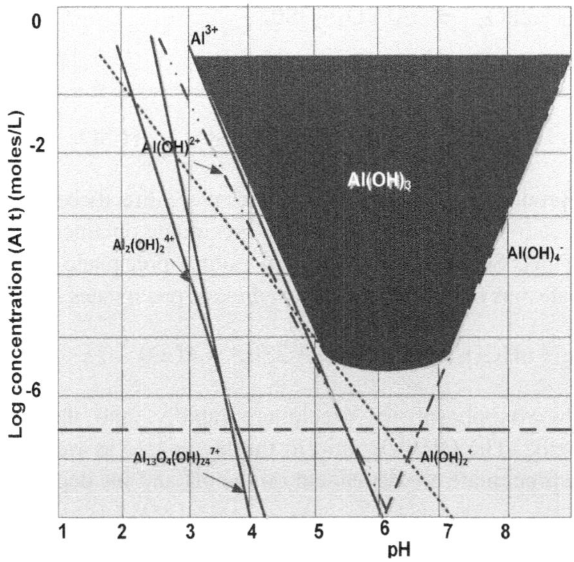

Figure 2.22. *Aluminum solubility as a function of pH (25°C)*

Chemical reaction of aluminum sulfate with phosphates

The basic reaction can be written as:

$$Al^{3+} + 3\ OH^- \rightarrow Al(OH)_3 \qquad Ks_{Al(OH)3} = 10^{-33}$$

$$Al^{3+} + PO_4^{3-} \rightarrow AlPO_4$$

or even as:

$$Al^{3+} + Hn\ PO_4^{3-n} \rightarrow AlPO_4 + nH^+$$

$Ks_{Al(OH)3} = 10^{-33}$ can be compared with $Ks_{AlPO4} = 10^{-21}$. Based on the value of $Ks_{Al(OH)3} = 10^{-33}$, we can deduce the Al^{3+} concentration, which can precipitate $Al(OH)_3$: 2.10^{-9} mole.L^{-1}. The same calculation can be done with $Ks_{AlPO4} = 10^{-21}$; the Al^{3+} concentration which can co-precipitate $AlPO_4$ is = 3.10^{-11} mole.L^{-1}. These two concentrations are similar, indicating that $Al(OH)_3$ and $AlPO_4$ co-precipitate.

This precipitate has a maximum solubility of 0.01 mg at pH 6 and is less soluble than ferric iron phosphate. But generally, in wastewater, pH $\geq$ 6.5.

In practice, with the use of $Al_2(SO_4)_3 \cdot 18H_2O$, which is most commonly used in precipitation, the reaction is written as:

$$Al_2(SO_4)_3 + 2PO_4^{3-} \rightarrow 2Al\,PO_4 + 3SO_4^{2-}$$

or even as:

$$Al_2(SO_4)_3 + 2\,NaH_2PO_4 \rightarrow 2AlPO_4 + Na_2SO_4 + 2\,H_2SO_4$$

The reaction producing $AlPO_4$ (variscite) takes place directly between the phosphate ions and the Al^{3+} cations. The solubility of the precipitate obtained, $AlPO_4$ (variscite), is a function of pH. Some of the aluminum cations precipitate with hydroxide and hydrogen carbonate ions from water to form hydroxide precipitates $Al(OH)_3$:

$$Al_2(SO_4)_3 + 6HCO_3^- \rightarrow 2Al(OH)_3 + 3SO_4^{2-} + 6CO_2$$

Aluminum hydroxophosphates precipitate rapidly, and the floc contains an excess of hydroxide. The OH/PO_4 ratio in the precipitate as well as the adsorptive properties of the precipitate are dependent on the pH and the degree of hydrolysis of Al ions.

Precipitation yields therefore depend on the Al/P ratio.

Phosphorus removal efficiency	**Al/P**
75%	1.2
85%	1.5
95%	2

Table 2.2. *Dephosphatation efficiency as a function of the Al/P ratio*

Product	Aluminum sulfate $Al_2(SO_4)_3$
Form	Liquid
% Al in the product	8–17
Density kg.L^{-1}	1.2–1.3
Precipitate formed	AlO_4
Stoichiometric molar ratio Al/P	1
Stoichiometric mass ratio Al/P	0.87
1 g of Al produced	4.52 g of $AlPO_4$ or 2.89 g of $Al(OH)_3$
1 g of P eliminated	3.94 g of $AlPO_4$

Table 2.3. *Characteristics of aluminum sulfate*

P yield	Al/P molar ratio	Al/P mass ratio
75%	1.4	1.2
85%	1.7	1.5
95%	2.3	2

Table 2.4. *P removal efficiency as a function of the Al/P ratio*

The phosphorus removal efficiency depends on various factors. These include the type of reagent used, the molar ratio, the sludge pH and the initial phosphorus concentration.

As an example, Table 2.4 shows the influence of the molar ratio on the efficiency of phosphorus removal at a wastewater treatment plant using an aluminum-based reagent.

The parameters we need to know are the following:

– Basicity indicates the number of hydroxyl ions included in the structure of a polymerized coagulant. The higher the basicity, the lower the impact on pH. Aluminum sulfate is not polymerized and therefore has no OH ions in its structure and no basicity.

– The Al_2O_3% for aluminum-based coagulants. For example, aluminum sulfate ($Al_2(SO_4)_3,14H_2O$) is in liquid form and generally contains 8.5% by weight of Al_2O_3, which corresponds to 4.6% by weight of aluminum (Al).

– The density, which is often expressed in kg.L^{-1}. For example, $Al_2(SO_4)_3 \cdot 14H_2O$, liquid (49% p/p) typically has a density of 1.32.

Calculating the coagulant concentration (mg.L^{-1}) is done as follows. Consider an aluminum-based coagulant $Al_2(SO_4)_3 \cdot 14H_2O$ (594 g.$mole^{-1}$) with an Al_2O_3 content of 8.5% and density of 1.32.

This is equivalent to: 594 x 8.5/100 = 50.5% in mass/mass of $Al_2(SO_4)_3 \cdot 14H_2O$.

The concentration of the coagulant is therefore: 10,000 x 50.5 x 1.32 = 666,000 mg.L^{-1} or 666 g.L^{-1}.

If the coagulant is calculated using a flow rate of 150 ml.min^{-1}, and if the untreated water flow rate is 50 L.s^{-1}, the coagulant concentration is: (150 x 666,000/(1,000 x 60))/50 = 33.3 mg.L^{-1}.

The precipitation of phosphate ions by aluminum salts can also be carried out by adding sodium aluminate ($Na_2OAl_2O_3$) or alum ($Al_2(SO_4)_3 \cdot 14H_2O$). The solid formed is composed of aluminum hydroxide ($Al(OH)_3$) and aluminum phosphate ($AlPO_4$), whose solubility products are respectively 3.0 10^{-34} and 9.84 10^{-21}. Given the acid–base balances of the different species in solution, precipitation is optimal at pH values between 6 and 6.5 (Lewandowski et al. 2003; Deronzier and Choubert 2004) and the aluminum phosphate precipitate is obtained according to the following equation:

$$Al_2(SO_4)_3 14H_2O + 2PO_4^{3-} \rightarrow 2AlPO_4(s) + 3SO_4^{2-} + 14H_2O$$

2.7.1.1.2. Iron-based coagulants

The main iron salts used in coagulation are ferric chloride, ferric sulfate, ferric chlorosulfate and ferrous sulfate. Ferric salts, being more highly charged, have a better coagulating power than ferrous salts.

The mode of action of iron salts, including ferric chloride, is generally quite well understood. In acidic solution, the ion Fe^{3+} is hydrated by six water molecules and forms an octahedral ligand. Depending on the pH, successive hydrolysis reactions occur:

$$Fe(H_2O)_6^{3+} + H_2O \rightarrow [Fe(H_2O)_5OH]^{2+} + H_3O^+$$

$$[Fe(H_2O)_5OH]^{2+} + H_2O \rightarrow [Fe(H_2O)_4(OH)_2]^+ + H_3O^+$$

$$[Fe(H_2O)_4(OH)_2]^+ + H_2O \rightarrow Fe(H_2O)_3(OH)_3 + H_3O^+$$

$$Fe(H_2O)_3(OH)_3 + H_2O \rightarrow [Fe(H_2O)_2(OH)_4]^- + H_3O^+$$

These successive reactions form iron hydroxyl complexes and cause iron to precipitate as a relatively stable floc. These reactions also lower the pH through the release of protons. The metal ions are linked by hydroxide bonds, and the hydrolysis products of the ferric ion neutralize the negative charges of the colloids, thus destabilizing them. Furthermore, the formation of polymerized compounds can produce large amounts of material that can then surround the colloidal particles, trapping them in the form of iron hydroxide. The isoelectric point of the system is around pH 8: below 8, the positively charged form predominates, while the negatively charged species is dominant when the pH is above 8. A high ferric salt concentration often induces rust-colored discoloration of the treated water: this is the main drawback of these products. The optimum pH range for this precipitation is between 5.5 and 8.5.

Ferric chloride ($FeCl_3, 6H_2O$)

Ferric chloride is available in liquid or solid form (crystalline or anhydrous). It is obtained by dissolving iron scrap in hydrochloric acid (formation of $FeCl_2$) followed by oxidation with liquid chlorine. It is commonly used in liquid form with a concentration of approximately 40–42% of $FeCl_3$ by weight. It also exists in crystalline form with a 60% content of $FeCl_3$ by weight. It is highly corrosive. Aqueous solutions of ferric chloride are very acidic and react with the alkali carbonates present in water. This reaction results in the release of carbon dioxide and a shift in the calcium carbonate equilibrium.

In addition to its coagulating action on colloidal substances, ferric chloride allows for the removal of certain soluble organic compounds. However, the optimal pH for the removal of organic matter is different and more acidic than the optimal pH required for turbidity reduction.

Fe^{3+} ions can be hydrolyzed and precipitated as $Fe(OH)_3$ at pH levels higher than 4.5. Good coagulation occurs within the pH range 5.5–8.5. The solubility product of $Fe(OH)_3$ is: $[Fe^{3+}]\,[OH^-]^3 = 10^{-38}$.

Unlike aluminum hydroxides, ferric hydroxide does not redissolve in a basic medium, which allows it to function at basic pH levels for certain applications.

In drinking water, these coagulants are used to remove color and organic matter with pH conditions below 5. As they often contain Mn^{2+}, the increased concentration of this metal in flocculated water must be taken into account by removing its particulate form during filtration. $FeCL_3$ coagulants are now available that contain less than 50 mg Mn^{2+}/kg $FeCL_3$ compared to the classic ratio of 600 mg Mn^{2+}/kg $FeCl_3$.

The low solubility of iron hydroxides (up to pH 8.5) leads to their being preferred over Al salt-based coagulants. However, they remain less effective in cold water than aluminum-based coagulants. Finally, $Fe(OH)_3$ flocs are denser and more voluminous than those of $Al(OH)_3$.

$FeCl_3$ reactions with alkalinity or lime are the following:

$$2\ FeCl_3 + 3\ Ca(HCO_3)_2 \rightarrow 2\ Fe(OH)_3 + 3\ CaCl_2 + 6\ CO_2$$

$$2\ FeCl_3 + 3\ Ca(OH)_2 \rightarrow 2\ Fe\ (OH)_3 + 3\ CaCl_2$$

Thus, 1 mg of ferric chloride consumes 0.92 mg of alkalinity and generates 0.66 mg of sludge.

Typical ferric chloride concentrations range from 5 to 190 $mg.L^{-1}$, depending on multiple factors, including the concentrations of natural organic matter (NOM) and the quality of the untreated water.

Previous studies have shown that the optimal pH range for iron-based coagulants is between 5 and 8.5, which is slightly higher than that of aluminum sulfate (Amirtharajah 1982).

However, determining the optimal pH range and coagulant concentration ultimately depends on the characteristics of the untreated water and differs from one water source to another.

Ferric sulfate ($Fe_2(SO_4)_3$, nH_2O)

Ferric sulfate is obtained from the reaction between ferric hydroxide and sulfuric acid. The number of hydrations is generally two or three, although some forms can have up to nine water molecules per ferric sulfate molecule. It is available in solid form (with 72–75% kg/kg $Fe_2(SO_4)_3$ and 20–21% Fe^{3+}) or in liquid form (with 40–42% kg/kg $Fe_2(SO_4)_3$ and 11.5% Fe^{3+}). However, this type of liquid coagulant may offer a much lower purity (30%). It is also corrosive, just like ferric chloride.

Like most ferric coagulants, it can be used in a pH range of 4 to 8.5 for drinking water applications. Its mode of action is similar to that of ferric chloride:

$$Fe_2(SO4)_3 + 9H_2O \rightarrow 2\ Fe(OH)_3 + 6\ H^+ + 3\ SO_4^{2-} + 3H_2O$$

The reactions of ferric sulfate with bases are as follows:

$$Fe_2(SO4)_3 + 3\ Ca(HCO_3)_2 \rightarrow Fe(OH)_3 + 3\ CaSO_4 + 6\ CO_2$$

$$Fe_2(SO4)_3 + 6\ NaHCO_3 \rightarrow Fe(OH)_3 + 3\ Na_2S0_4 + 6\ CO_2$$

$$Fe_2(SO4)_3 + 3\ Na_2CO_3 + 3H_2O \rightarrow 2\ Fe(OH)_3 + 6\ CO_2$$

$$Fe_2(SO4)_3 + 6\ NaOH \rightarrow 2\ Fe(OH)_3 + 3\ Na_2SO_4$$

$$Fe_2(SO4)_3 + 3\ Ca(OH)_2 \rightarrow 2\ Fe(OH)_3 + 3\ CaSO_4$$

Thus, 1 mg of ferric sulfate consumes 0.75 mg of alkalinity and generates 0.54 mg of sludge in the form of iron hydroxide. Ferric sulfate behaves somewhat like ferric chloride but is less commonly used in wastewater treatment plants. Typical concentrations vary from 10 to 250 $mg.L^{-1}$.

Ferrous sulfate ($FeSO_4$, $7H_2O$)

Ferrous sulfate is the product of the reaction between sulfuric acid and iron. This product is subject to certain constraints, including the requirement to maintain the solution above 10°C. It exists in solid or liquid form. It is most often supplied in the form of crystals or granules containing approximately 20% of Fe. Ferrous sulfate must be oxidized to $Fe(OH)_3$ to be usable in coagulation. Ferrous sulfate is used optimally at high values of pH > 7.7. The combination of ferrous sulfate with chlorine is carried out just upstream of the coagulation step to produce ferric chloride and ferric sulfate according to the following simplified reaction:

$$6\ Fe(H_2O)_7SO_4 + 3\ Cl_2 \rightarrow Fe_2(SO_4)_3 + 2\ FeCl_3 + 42\ H_2O$$

The ferric chloride and ferric sulfate molecules react in turn and cause coagulation.

The reactions of $FeSO_4\ 7H_2O$ with alkalinity or lime are the following:

$$FeSO_4\ 7\ H_2O + Ca(OH)_2 \rightarrow Fe(OH)_2 + CaSO_4 + 7\ H2O$$

$$4Fe(OH)_2 + O_2 + 2H_2O \rightarrow 4Fe(OH)_3$$

The solubility product of hydroxide Fe^{2+} is: $Ks = 10^{-15,1}$.

Under certain conditions, there will initially be an oxidation reaction of Fe^{2+} into Fe^{3+} followed by the precipitation of Fe^{3+} with phosphate ions.

Ferric chlorosulfate ($FeClSO_4$)

Ferric chlorosulfate is commercially available in liquid form. This product is obtained by the oxidation of ferrous sulfate with chlorine, or by the reaction of acids with ferric oxides:

$$6FeSO_4,\ 7H_2O + 3Cl_2 \rightarrow 2Fe_2(SO_4)_3 + 2FeCl_3 + 7H_2O$$

The combination of ferrous sulfate with chlorine is carried out just upstream of the coagulation step to produce ferric chloride and ferric sulfate. Thus, 1 $mg.L^{-1}$ of $FeSO_4$ theoretically requires 0.13 mg/l chlorine.

	Formula	**Fe content**	**Density at 20°C**	**pH at 20°C**	**Molecular weight in g/mole**
$FeCl_3$ Ferric chloride	$FeCl_3$, $6H_2O$	14–15%	1.44	1.8	162.5
$FeSO_4$ Ferrous sulfate	$FeSO_4$,$7H_2O$	20%	0.9	1.5–2.2	152
$Fe_2(SO_4)_3$ Ferric sulfate	$Fe_2(SO_4)_3\ 9H_2O$	12–25%	1.44–1.55	< 2	400
PFS Polyferric sulfate	$Fe_2(OH)_{0.6}$ $(SO_4)_{2.7}$	12.2%	1.50	< 2	382.4

Table 2.5. *Physicochemical characteristics of Fe-based coagulants*

Ferric salts, when in solution, dissociate immediately to form a hydrated reaction of the products. The metal ions form coordination compounds with water molecules to give $\{Fe(H2O)_6\}^{3+}$.

The equation provides the overall stoichiometric reaction of iron and shows that for each mole of trivalent metal ion, one mole of amorphous solid (am) and three moles of hydrogen ions are produced. Iron acts as an acid because it releases hydrogen ions and lowers the pH of water:

$$Fe^{3+} + 3H_2O \rightarrow Fe(OH)_3 + 3\ H^+$$

A similar reaction occurs with aluminum sulfate.

The pH range for iron precipitation is approximately 5 to 8, and its minimum solubility occurs at a pH of approximately 8. Ferric species are less soluble over a wider pH range (Letterman and Yiacoumi 2011). Positively charged hydrolysis products and Fe^{3+} ions form at low pH levels, while only negatively charged hydrolysis products form at higher pH levels.

One observation from the solubility diagrams is that iron-based coagulants have a wider pH range, within which precipitation is likely to occur, compared to aluminum-based coagulants. The curves surrounding the iron hydroxide precipitate zone indicate the equilibrium concentrations of the soluble species in equilibrium with $Fe(OH)_3(s)$. For a given concentration of coagulant (expressed in Fe^{3+}), these curves delimit the pH range where the products of ferric chloride hydrolysis are most effective for coagulation and destabilization of colloids.

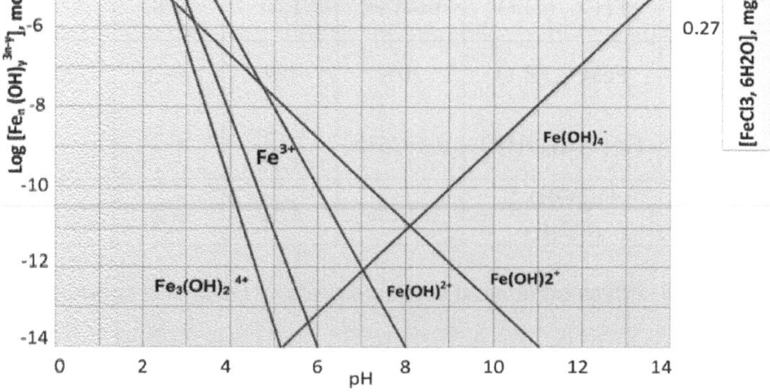

Figure 2.23. *Iron hydroxide solubility equilibrium diagram at 25 °C (Crittenden et al. 2005)*

The reaction of Fe^{3+} with phosphates is written:

$$Fe^{3+} + PO_4 H^{2-} \rightarrow FePO_4 + 2 H^+$$

The dominant chemical reactions between ferric ions and phosphates are:

$$FeCl_3 + NaH_2PO_4 \rightarrow FePO_4 + NaCl + 2 HCl$$

$$FeSO_4Cl + NaH_2PO_4 \rightarrow FePO_4 + NaCl + H_2SO_4$$

Precipitation is dependent on pH and the contact time between iron ions and phosphates. In the presence of phosphate ions, Fe^{3+} reacts to form a precipitate. Phosphate anion precipitation is optimal at pH values between 6 and 6.5. The iron salts generally used are ferric chloride ($FeCl_3$) and ferrous sulfate ($FeSO_4$) salts (Deronzier and Choubert 2004). Iron chloride salt (III) ($FeCl_3$) reacts with the phosphate anion according to the following equation:

$$FeCl_3(s) + PO_4^{3-} \rightarrow FePO_4 + 3Cl^-$$

The solubility of the precipitate $FePO_4$ (strengite) obtained is pH-dependent. The optimal pH is in the range of 6–6.5. The added ferric ions precipitate in parallel with the hydroxide and carbonate ions in the water to form iron hydroxide precipitates:

$$Fe^{3+} + 3 OH^- \rightarrow Fe(OH)_3$$

$$Fe^{3+} + 3 HCO_3^- \rightarrow Fe(OH)_3 + 3 CO_2$$

Thus, while in theory one mole of Fe is needed to precipitate one mole of P, in practice, the molar ratio to apply Fe/P is higher to account for these "parasitic" reactions. The molar ratio depends on the initial phosphorus concentration and the desired phosphorus removal efficiency.

The quantities of reactant used and the quantities of precipitate formed can be deduced from the chemical equations presented below.

For 1 mole of iron (= 56 g of iron) produced:

– 1 mole of $FePO_4$ (51 g of $FePO_4$).

– 1 mole of $Fe(OH)_3$ (107 g of $Fe(OH)_3$).

This can also be written as: for 1 g of iron produced:

– 151/56 = 2.7 g of $FePO_4$.

– 107/56 = 1.9 g of $Fe(OH)_3$.

It can also be written as: for 1 g of P eliminated product, 151/31 = 4.87 g of $FePO_4$.

This reaction produces insoluble iron phosphates (see Table 2.7) such as strengite ($FePO_4$) with a solubility product Ks = $9.91.10^{-16}$ or vivianite ($Fe_3(PO_4)_2$) with a solubility product Ks = 10^{-36} (Stumm and Morgan 1996; De-Bashan et al. 2004).

Ferrous sulfate ($FeSO_4$)

Ferrous sulfate ($FeSO_4$) is a rarely used reagent. Depending on the distributor, it is offered in either liquid form (iron representing 12.5% by weight) or powder form. When ferrous sulfate is added to the aeration tank, by far the most frequent case, the ferrous ions are oxidized to ferric ions very rapidly. Phosphorus therefore precipitates with the ferric ions according to the reaction described previously. The remaining ferrous ions lead to the formation of the precipitate $Fe_3(PO_4)_2$ (or vivianite) according to the following reaction:

$$3\ FeSO_4 + 2\ NaH_2PO_4 \rightarrow Fe_3(PO_4)_2 + Na_2SO_4 + 2\ H_2SO_4$$

Table 2.6 shows the quantities of commercial reagent to be injected to remove one tone of phosphorus, taking an Fe/P molar ratio of 1.

Product	Ferric chloride $FeCl_3$	Ferrous sulfate $FeSO_4$
Form	Liquid	Liquid/powder
% Fe in the product	14	12.5 (liquid)
Density kg.L^{-1}	1.4–1.5	–
Precipitates formed	$FePO_4$; $Fe(OH)_3$	$Fe_3(PO_4)_2$ $Fe\ (OH)_3$
Stoichiometric molar ratio Fe/P	1	1.5
Stoichiometric mass ratio Fe/P	1.81	2.71
1 g of Fe produced	2.7 g of $FePO_4$ or 1.9 g of $Fe\ (OH)_3$	2.7 g of $FePO_4$ or 1.9 g of $Fe(OH)_3$ or 4.26 g of $Fe_3(PO_4)_2$
1 g of P eliminated product	4.857 g of $FePO_4$	

Table 2.6. *Characteristics of ferric chloride and ferrous sulfate*

Precipitate	Chemical formula
Strengite	$FePO_4H_2O$
	$Fe_{2,5}PO_4(OH)_{4,5}$
	$Fe_{1,6}H_2PO_4(OH)_{3,8}$
Vivianite	$Fe_3(PO_4)_2$

Table 2.7. *Main iron phosphate compounds (Comeau et al. 2006)*

2.7.1.1.3. Lime reagent

This reagent is reminiscent of an old wastewater treatment practice. Its application in the precipitation treatment of soluble phosphates involves the formation of insoluble hydroxyapatite ($Ca_5(OH)PO_4)$) in the presence of OH^- ions:

$$3\ HPO_4^{2-} + 5\ Ca^{2+} + 4\ OH^- \rightarrow Ca_5(OH)PO_4)_3 + 3\ H_2O$$

$$Ca_5(OH)PO_4)_3 \rightarrow 5\ Ca^{2+} + 4\ OH^- + 3\ PO_4^{3-}$$

$$Ks = 10^{-55.9}$$

The theoretical molar ratio Ca/P is 1.67.

The chemical reactions are as follows:

$$Ca^{2+} + HCO_3^- + OH^- \rightarrow CaCO_3 + H_2O$$

$$3\ H_2PO_4^- + 5\ Ca^{2+} + 7\ OH^- \rightarrow Ca_5(OH)PO_4)_3 + 6\ H_2O$$

$$Mg^{2+} + 2\ OH^- \rightarrow Mg\ (OH)_2$$

Excess calcium can be recarbonated:

$$Ca^{2+} + CO_2 + 2OH^- \rightarrow CaCO_3 + H_2O$$

or precipitated as $CaCO_3$:

$$Ca^{2+} + CO_3^{2-} + 2Na^+ \rightarrow CaCO_3 + 2Na^+$$

The solubility of hydroxyapatite decreases with increasing pH, and consequently, phosphorus removal increases with pH. At a pH above 9.5, most of the hydroxyapatite is insoluble.

During "natural" precipitation (precipitation of phosphates without the addition of reactants), the main precipitated phosphate compounds are also found in the form of hydroxyapatite ($Ca_5(OH)PO_4)_3$) or struvite ($MgNH_4PO_4$), and vivianite ($Fe_3(PO_4)_2$).

In the case of biological dephosphatation, these precipitates form in the anaerobic tank because this is where the concentrations of calcium and magnesium ions are highest. The presence of organic matter or polyphosphates does not play a role.

There are many chemical reactions between Ca^{2+} and P, depending on the chemical form of the phosphates. Some reactions are summarized below:

$$Ca^{2+} + H_2PO_4^- \rightarrow CaH_2PO_4 \qquad pK = -1.08$$

$$Ca^{2+} + HPO_4^{2-} \rightarrow CaHPO_4 \qquad pK = -7.0$$

$$3Ca^{2+} + 2PO_4^{3-} \rightarrow Ca_3(PO_{4)2} \qquad pK = -26$$

$$Ca^{2+} + HPO_4^{2-} + HCO_3^- \rightarrow Ca_2HPO_4CO_3 + H^+ \qquad pK = 1.33$$

$$Ca_2HPO_4CO_3 \rightarrow Ca_2PO_4CO_{3-} + H^+ \qquad pK = 8.3$$

$$10CaCO_3 + 6HPO_4^{2-} + 2H_2O + 2H^+ \rightarrow Ca_{10}(PO_4)_6(OH)_2 + 10HCO_3^- + H^+ \qquad pK = -32$$

Some forms of calcium phosphate precipitates are reported in Table 2.8. The precipitation reaction of phosphate anions and calcium depends on the molar ratio Ca/P (see Table 2.8).

Precipitates	Abbreviation	Chemical formula	Ca/P (mole/mole)
Hydroxyapatite	HAP	$Ca_5(PO_4)_3(OH)_2$	1.67
Fluorapatite	FAP	$Ca_{10}(PO_4)_6F_2$	1.67
Octocalcium phosphate	OCP	$Ca_4H(PO_4)_3$	1.33
Tricalcium phosphate	TCP	$Ca_3(PO_4)_2$	1.5
Dicalcium phosphate	DCP	$Ca_2(HPO_4)$	1
Monocalcium phosphate	ACP	$Ca(H_2PO_4)_2$	0.5

Table 2.8. *Main calcium phosphate precipitates (Cabanes 2006)*

Indeed, for an initial solution Ca/P ratio of approximately 1.67, homogeneous precipitation becomes effective at a pH of 8.5–9. Other parameters such as temperature, ionic strength, pH and calcium and phosphorus concentrations impact the precipitation of calcium phosphates. Hydroxyapatite and fluorapatite are generally the main calcium phosphates formed, with hydroxyapatite being the most stable calcium phosphate ($Ca_5(PO_4)_3(OH)$) with solubility product Ks = $2.07.10^{-33}$ (Cabanes 2006).

Direct action of lime on orthophosphates

The following equation describes the formation of hydroxyapatite after precipitation of the phosphate anion by the calcium ion:

$$3PO_4^{3-} + 5Ca^{2+} + OH^- \rightarrow Ca_5(PO_4)_3(OH)(s)$$

Hydroxyapatite formation is observed under neutral pH conditions in natural waters, in the presence of bacteria and for minimum phosphorus concentrations between 50 mg L^{-1} and 100 mg L^{-1} of calcium.

Adsorption of orthophosphates onto calcium carbonate formed by the action of lime on calcium bicarbonate

$$3Ca(HCO_3)_2 + 3Ca(OH)_2 \rightarrow 6CaCO_3 + 6H_2O$$

$$4H_3PO_4 + 6CaCO_3 \rightarrow 2Ca_3(PO_4)_2 + 6H_2CO_3$$

The pH of the raw water affects the efficiency of the reaction. A pH of 9.5 is necessary, and a specific mass ratio is required ($Ca(OH)_2/PO_4$ equal to 70) for good phosphate removal (>90%). The advantage of potentially using lime is that the resulting sludge allows for lime recovery after calcination.

2.7.1.1.4. Magnesium salts

Magnesium salts are not widely used for precipitating phosphate ions due to their high cost. However, these salts are valuable because, in addition to phosphate ions, they also remove ammoniacal nitrogen and magnesium (Lee et al. 2003) according to the following equation:

$$Mg^{2+} + NH_4^+ + 6H_2O + PO_4^{3-} \rightarrow MgNH_4PO_4{,}6H_2O(s)$$

A white crystalline compound made up of magnesium and ammonium phosphate called struvite is formed ($MgNH_4PO_4{,}6H_2O$) (Battistoni et al. 2000; Quintana et al. 2008).

In summary, the chemical precipitation process involves the continuous control of physicochemical parameters (pH, phosphate ion concentration, alkalinity, background salt concentration and temperature) that generate the dissolution–precipitation equilibria of phosphate salts. Furthermore, the disposal and treatment of precipitate sludge is costly. This process is widely used and well suited for large municipalities but is too expensive for small municipalities and developing countries.

2.8. Implementation of chemical reagents

The most common chemical method for removing phosphorus (P) involves introducing metallic salts into pretreated influencers, CAS reactors or secondary clarification effluents. The added reagents precipitate the P in the wastewater, and the resulting precipitates are removed by sedimentation, flotation or rotary sieves. It all depends on where in the treatment process phosphate removal takes place. Phosphate removal rates are generally proportional to the mass of chemical added, which impacts the amount of solids produced that must be added to the biological sludge production.

2.8.1. *Injection points*

Metal ions must be injected into a well-agitated zone to ensure proper reagent dispersion and prevent any bypass into the treated effluent. The injection point should be optimized at startup, especially if physicochemical removal is combined with biological treatment. When the mixture is correct, reagent consumption should remain constant.

The recommendations for injection sites are:

– at the exit of an aerobic zone or within the aerobic zone located downstream;

– in the aerated zone;

– two thirds into the tank if there is a piston flow.

The establishment of a minimum reagent line length must be considered when choosing the injection point.

2.8.2. *Injection control*

The reagent will be injected at a constant flow rate with sequential operation of the feed pumps. The start-up rate and pumping time will be adjusted according to the average flow rate to be treated and will take into account the residual amount obtained at the outlet.

Control based on specifically measured phosphate concentrations or direct control of the water flow rate to be treated is possible but expensive.

Unlike the biological route, the physicochemical route offers some flexibility with regard to the injection site: it is possible to inject during primary treatment (pre-precipitation), secondary treatment (simultaneous precipitation) or tertiary treatment (post-precipitation).

In practice, ferric ions also react with hydroxide ions (OH^-) and carbonates (HCO_3) to form iron hydroxide precipitates.

These reactions are called “parasitic reactions”:

$$Fe^{3+} + 3\ OH^- \rightarrow Fe(OH)_3$$

$$Fe^{3+} + 3\ HCO_3^- \rightarrow Fe(OH)_3 + 3\ CO_2$$

These parasitic reactions tend to consume alkalinity, sometimes inducing a sharp drop in pH (acidified effluents) under certain effluent quality conditions (weakly buffered effluents, low carbonate concentration).

The $FeCl_3$ concentration is therefore generally higher due to competition between the different reactions (Besnault et al. 2023). The molar ratio must therefore be greater than “Fe/P = 1/1” and adapted to each context of chemical phosphorus treatment implementation. Jar tests can be carried out to best define the feasible concentrations. Physicochemical phosphorus removal therefore also leads to the production of mineral sludge due to the production of iron phosphate precipitates and iron hydroxides. In activated sludge, the average overproduction of physicochemical sludge induced is around 20% in the case of physicochemical phosphorus removal (Deronzier and Choubert 2004).

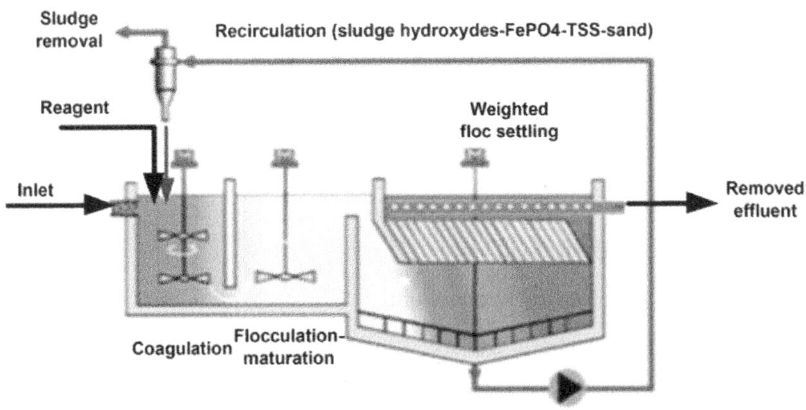

Figure 2.24. *Weighted floc settling (Actiflo® Veolia) in primary settling tank or tertiary treatment*

The addition of specific concentrations of chemicals is generally reliable and widely accepted; therefore, it is the most commonly used phosphorus treatment option.

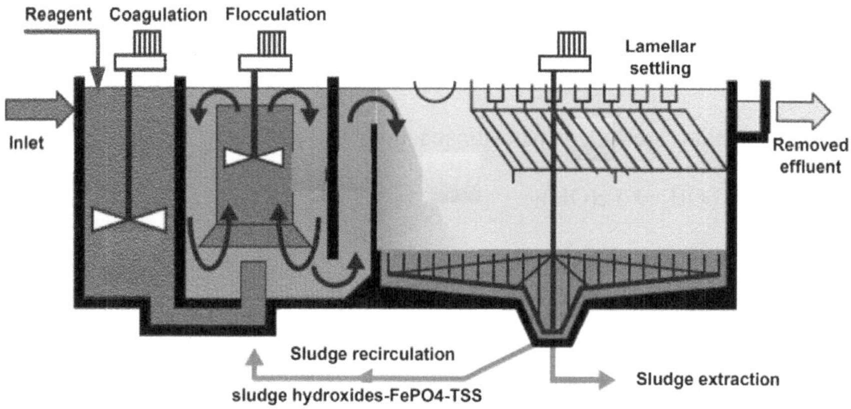

Figure 2.25. *Lamellar settling (Densadeg®) in primary settling or tertiary treatment tank*

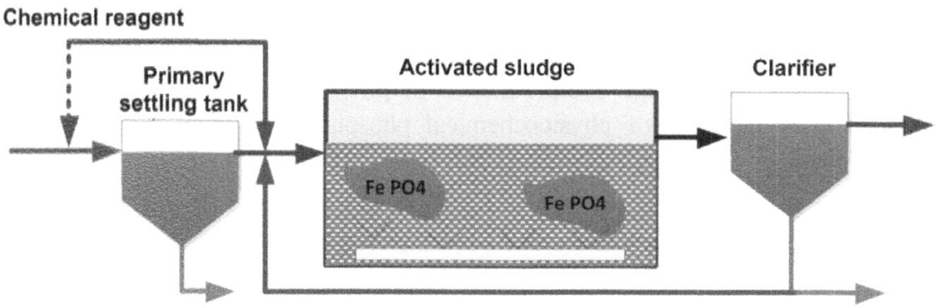

Figure 2.26. *Simultaneous physicochemical dephosphatation with advanced biology*

Phosphating can be performed via:

– primary settling with reagent injection;

– bacterial assimilation (ratio BOD/P: 100/1);

– anaerobic biological reactions according to several configurations;

– injecting reagent into the biological tank;

– clarification for further refinement.

In direct precipitation, the mineral coagulant is applied upstream of the primary settling tank, which does not necessarily mean that no further treatment is required. It all depends on the phosphorus consumption by the biomass and the desired final concentration.

This process is used when the receiving environment is only slightly affected by relatively high concentrations of organic matter and nitrogen, but due to the risk of eutrophication, very efficient phosphorus removal is necessary. Phosphorus precipitation is accompanied by significant removal through the formation of a precipitate ($FePO_4$ or $AlPO_4$), and the subsequent biological treatment operates with a lower TSS and BOD load. However, only a small amount of organic phosphorus is removed during this precipitation stage. Furthermore, colloids and other ionic species present in the untreated water consume reagent, which must therefore be introduced in excess to ensure proper removal. Sludge volumes are 50 to 100% greater than the volumes produced without the addition of reagents. A reduction of 90% is achieved for phosphorus removal, and the phosphorus concentration is less than 1 $mg.L^{-1}$.

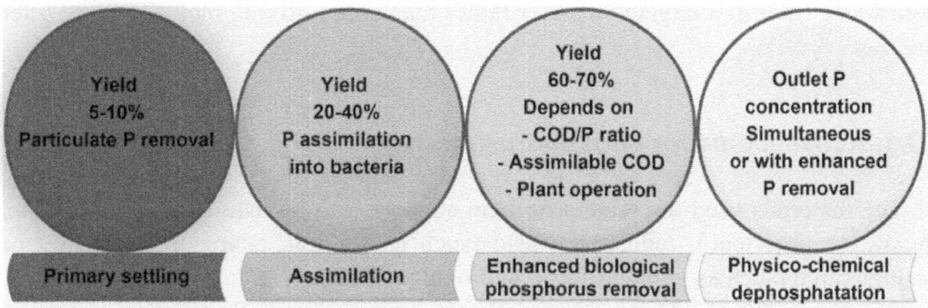

Figure 2.27. *Different stages of phosphorus removal in a wastewater treatment plant*

Simultaneous precipitation (co-precipitation) is suitable for activated sludge treatment plants, where the reagents are introduced directly into the aeration tank. The coagulation–flocculation process and phosphorus assimilation by the biomass allow for a 60 to 70% reduction in chemical consumption. This configuration allows the elimination of organic phosphorus oxidized to phosphates in the biological reactor. A 40 to 65% increase in sludge volume can occur compared to a solution without the addition of reagents. Post-precipitation is a refining treatment performed on the clarified effluent. Small quantities of metallic reagents are added. This process allows for very good phosphorus removal, and efficiency can reach 95%. The P concentration in the effluent can be less than 1.3 mg. L^{-1}. The process reduces residual (colloidal) suspended solids after clarification. Chemical reagents should preferably be injected during strong agitation to ensure good dispersion and guarantee rapid precipitate formation. This step is followed by one with medium agitation, allowing flocculation while avoiding excessive mixing that would destroy the flocs.

Depending on the configuration, the main characteristics to consider are the chemical reagents and their implementation, the required concentration, the injection point, the coagulation and flocculation conditions, the equipment required, the control method, the solid–liquid separation and the sludge produced. The addition of aluminum sulfate causes a decrease in pH and a drop in alkalinity. The addition of ferric chloride causes a decrease in pH, a drop in alkalinity and greater sludge production than with aluminum salt.

2.9. Other technical solutions for phosphorus removal

Improvements in phosphorus removal in active media filtration systems have recently attracted attention. These filters, unlike conventional filtration methods, rely on the phosphorus sorption capabilities of certain materials to remove phosphorus from wastewater in a targeted manner rather than simply fixing biomass using filter media.

2.9.1. *Active media for phosphorus removal*

The materials used to create adsorbent supports can be natural (such as apatite, bauxite or limestone), industrial waste (such as fly ash, ochre or steel slag) or manufactured (such as granular ferric hydroxide (GFH)).

Phosphorus is removed by sorption or direct precipitation. This involves the movement of phosphorus from the wastewater to the surface of the active support.

A large number of substrates have been studied and implemented on a small scale. They all share the characteristic of containing elements such as iron, calcium, aluminum or magnesium in their chemical composition.

2.9.1.1. *Natural materials*

Numerous natural materials have been studied for the adsorption of PO_4^{3-} ions (see Table 2.9). Opoka (a Polish rock composed of silica and calcium carbonate) has been used as an adsorbent for the dephosphatation of water containing 10 mg L^{-1} P-PO_4 and gave a maximum adsorption capacity (qmax) of 0.1 mg g^{-1}. Bauxite, limestone and zeolite have a maximum adsorption capacity between 0.25 and 0.6 mg g^{-1} (Drizo et al. 1999). Apatite, a phosphate rock, has also been studied in a column reactor and as a material for artificial marshes (Molle et al. 2005; Bellier et al. 2006). Natural apatite is a calcareous material with significant potential for phosphate removal.

It has been shown that a strong chemical affinity exists between apatite and phosphorus; it is a material composed of 37% calcium and 17% phosphorus by mass percentage. This material has a relatively high phosphate anion adsorption capacity close to 4.76 mg g^{-1}. A product of marine origin (sea maerl) also showed good retention capacity (q_{max} = 39.5 mg g^{-1}) (Gray et al. 2000). However, all these experiments were conducted with high phosphate concentrations, which can affect adsorption capacity. These values do not reflect reality, and the selected materials should be tested with the wastewater concentrations leaving the treatment plant. With the polonite material, a phosphate removal rate of 91% was achieved over a one-year period with a P sorption capacity of 120 g.kg^{-1} (Renman and Renman 2010).

2.9.1.2. *Synthetic materials*

Several studies have highlighted the effectiveness of synthetic products in removing phosphate ions. The use of GFH and activated aluminum oxide to remove phosphorus is a simple solution to implement. GFH has demonstrated an adsorption capacity and affinity for phosphorus, allowing for very low outlet concentrations. The support is regenerated using sodium hydroxide.

Biochar composites containing magnesium oxide nanoparticles (MgO-biochar) have been synthesized, exhibiting significant P adsorption capacities attributed to their external surface, with a maximum capacity of 60.95 mg P g^{-1} (Zhu et al. 2020).

Zirconium-loaded okara has also proven effective for capturing P from aqueous solutions, with a maximum adsorption capacity of approximately 44.13 mg PO_4.g^{-1}, demonstrating rapid, feasible, spontaneous and endothermic elimination processes (Nguyen et al. 2014).

Furthermore, dolomite-modified biochar derived from dewatered urban sewage sludge showed a high phosphate removal efficiency of 96.8%, the adsorption process being endothermic and well described by the Langmuir model, indicating electrostatic attraction and the formation of new surface complexes as phosphate removal mechanisms.

Iron green rusts and ferrihydrite were among the species studied (Barthelemy et al. 2012; Mallet et al. 2013). Under acidic pH conditions (pH ~4), a maximum adsorption capacity for phosphate anions of 64.8 and 105 mg g^{-1} has been reported for ferric green rusts and for ferrihydrite, respectively, for an initial phosphate concentration of 100 mg L^{-1}. The adsorption capacity remains high for iron-ferrihydrite at neutral pH – which is representative of water treatment conditions – at 78 mg PO_4 g^{-1}. However, these materials must be deposited on a porous surface.

Material	P-PO_4 (mg L^{-1})	t (h)	R (%)	qmax (mg g^{-1})	pH	Authors
Opoka	10	20	20	0.1	8.3	Johansson (2000)
Bauxite	40	24	60	0.60	5.9	Drizo (1999)
Limestone	40	24	55	0.55	7.8	Drizo (1999)
Zeolite	40	24	25	0.25	6.6	Drizo (1999)
Dolomite	100	24	–	0.3	11	Pan et al. (2000)
Shell sand	480	24	67	9.6	8.8	Ádám et al. (2007)
Apatite	500	24	–	4.76	7	Molle (2005)
Maerl	5000	48	32	39.5	7.6	Gray (2000)

Table 2.9. *Experimental data relating to phosphate adsorption on natural materials (Vohla 2011). R: removal rate; qmax: maximum adsorption capacity; t: contact time*

However, operation at pH 4 requires neutralization of the treated effluent by adding sodium hydroxide. Supports such as pozzolans, expanded clay or iron filings have been successfully tested.

After saturation with phosphate ions, most of these synthetic adsorbents can be regenerated by desorption using basic aqueous solutions such as sodium hydroxide. However, the manufacturing cost of these synthetic materials, the fate of the regeneration concentrates and the conditions for cleaning and extracting the deposited flocs are drawbacks that operators must overcome.

2.9.1.3. *Industrial by-products*

Various industrial by-products, such as alumina industry waste (red mud), iron and steel by-products, and thermal incineration ash slag (oil shale ash, fly ash), have been evaluated for potential use in wastewater treatment (Bowden et al. 2009). The ash from bituminous shale was evaluated in work by Kaasik et al. (2008) for an initial phosphate ion concentration of 98 mg L^{-1} P-PO_4^{3-} and for a contact time of 48 h. The maximum adsorption capacity obtained was 6.5 mg g^{-1} P-PO_4^{3-}. Electric kiln slags were studied in a column reactor fed with synthetic effluent containing 500 mg L^{-1} of phosphate (Bowden et al. 2009) with a retention time of 24 h. The maximum adsorption capacity (qmax) obtained was 89.9 mg g^{-1} P-PO_4^{3-}. As for fly ash (Xu et al. 2006), a good affinity for phosphate anions was demonstrated. Indeed, for an initial concentration of 1,000 mg L^{-1} P.PO_4^{3-} and a contact time of 24 h, the maximum reported adsorption capacity was 8.5 mg g^{-1} P.PO_4^{3-}. As with the other tested substrates, the high concentrations of phosphates in contact with these substrates do not reflect real-world conditions; that is, concentrations lower than 10 mg L^{-1} P-PO_4^{3-}.

2.9.2. *Artificial wetlands*

Artificial wetlands are of interest because phosphorus (P) can be absorbed through microbial and plant activity, in addition to adsorption by the surrounding environment. Although P removal in wetlands is limited to 40–60% (Vymazal 2007), these systems are fairly well understood, and the flexibility of their configurations means they can suit a wide range of applications, being capable of simultaneously removing multiple contaminants.

The advantages of "adsorbent" wetlands are their low maintenance requirements, their aesthetic appeal and their ability to reduce BOD and nitrogen. It should be noted, however, that artificial wetlands often require a continuous water flow, which may limit their application to small-scale treatment plants. In some applications, adsorbent media have been incorporated into the filter medium, allowing for a complementary action between physicochemical mechanisms (adsorbent media) and biosorption (vegetation and bacteria).

This solution has some significant limitations. The main concern is its impact on pH and the costs associated with its rehabilitation (fieldwork, substrate replacement, removal and disposal of the substrate).

2.9.3. *Ion exchange*

Ion exchange technologies are well established and used in many contexts, such as water desalination and deionization. The principle of ion exchange can also be applied to phosphorus removal from wastewater. The highly selective nature of ion exchangers allows for the targeting of specific parameters. The predominant form of phosphorus in wastewater effluents is anionic. Phosphate ions are reversibly exchanged with a cation in the ion exchanger. There are resins with tertiary amines ($(N(CH_3)_2H)$ fixed to a macroporous styrene-divinylbenzene copolymer (Taiyuan Lanlang Technology Industrial Corp.). This resin is particularly suitable for removing phosphorus.

The performance of traditional ion exchangers is governed by the valence and weight of the ion in question, but due to the relatively low concentration of phosphate ions in wastewater, early attempts to implement this technology failed (Nesbitt 1969).

More recent attempts have improved efficiency through pretreatment of ion exchange media, for example, with ferric oxide or aluminum hydroxide, to increase selectivity for phosphate ions and have resulted in P removal rates of 80 to 90% (Martin et al. 2009; Seo et al. 2013).

Ion exchange systems have the advantage of allowing the recovery of phosphorus (P) by post-treatment of the sorption media. Although high rates of P removal have been achieved at the laboratory scale, large-scale implementation has been limited due to the need to add expensive chemicals for P recovery and the sensitivity of some media to pH conditions.

2.10. Overall phosphorus removal performance

Table 2.10 presents the phosphate removal performance under different configurations. For tertiary treatment, refer to the processes described in Volumes 1 and 2 of *Wastewater Reuse* (Gaid 2025a, 2025b).

Solutions implemented	P-total ($mg.L^{-1}$)	P-PO_4 ($mg.L^{-1}$)
Primary treatment with coagulation–flocculation–sedimentation	< 2.5	< 0.5
Co-precipitation in activated sludge	< 1.0	< 0.3
Primary chemical treatment and Biofilter (Biostyr)	< 1.0	< 0.3
Tertiary treatment		
Activated sludge/clarification including co-precipitation/Hydrotech sieve	< 0.6	< 0.3
Primary chemical treatment – Biofilter (Biostyr) – Hydrotech sieve	< 0.6	< 0.3
Activated sludge with co-precipitation and sand filtration	< 0.5	< 0.3
Primary chemical treatment – Biofilter (Biostyr) – sand filtration	< 0,5	< 0.3
Activated sludge with co-precipitation – Biofilter (Biostyr) – post-denitrification	< 0.5	< 0.2
Hydrotech sieve – sand filtration with coagulation	< 0.2	< 0.1
Activated sludge with co-precipitation – tertiary Actiflo®	< 0.3	< 0.1
Primary chemical treatment – Biofilter (Biostyr) – tertiary Actiflo®	< 0.3	< 0.1
Activated sludge with co-precipitation – tertiary Actiflo® – Hydrotech sieve	< 0.2	< 0.05
Primary chemical treatment – Biofilter (Biostyr) – tertiary Actiflo® – Hydrotech sieve	< 0.2	< 0.05
Activated sludge with co-precipitation – MBR	< 0.3	< 0.2
Activated sludge with co-precipitation – MBR-Actiflo®	< 0.1	< 0.02
Activated sludge or Biofilter (Biostyr) – microfiltration/reverse osmosis	< 0.02	< 0.02

Table 2.10. *Performance of some phosphate treatment configurations*

2.11. References

Acevedo, B., Oehmen, A., Carvalho, G., Seco, A., Borrás, L., Barat, R. (2012). Metabolic shift of polyphosphate-accumulating organisms with different levels of polyphosphate storage. *Water Res.*, 46, 1889–1900.

Acevedo, B., Borras, L., Oehmen, A., Barat, R. (2014). Modelling the metabolic shift of polyphosphate-accumulating organisms. *Water Res.*, 65, 235–244.

Ádám, K., Søvik, A.K., Krogstad, T., Heistad, A. (2007a). Phosphorus removal by the filter materials light-weight aggregates and shell sand – A review of processes and experimental setups for improved design of filter systems for wastewater treatment. *Vatten*, 63, 245–257.

Ádám, K., Krogstad, T., Vråle, L., Søvik, A.K., Jenssen, P.D. (2007b). Phosphorus retention in the filter materials shell sand and filtralite P® batch and column experiment with synthetic P solution and secondary wastewater. *Ecol. Eng.*, 29(2), 200–208.

Amirtharajah, A.M.K. (1982). Rapid-mix design for mechanisms of alum coagulation. *J. Am. Water Works Assoc.*, 74(4), 210–216.

Arias, C.A., Brix, H., Johansen, N.H. (2003). Phosphorus removal from municipal wastewater in an experimental two-stage vertical flow constructed wetland system equiped with a calcite filter. *Water Sci. Technol.*, 48, 51–58.

Arnz, P., Arnold, E., Wilderer, P.A. (2001). Enhanced biological phosphorus removal in a semi full-scale SBR. *Water Sci. Technol.*, 43, 167–174.

Arvin, E. (1985). Biological removal of phosphorus from wastewater. *Crit. Rev. Environ. Control*, 15(1), 25–64.

Barnard, J.L. (1975). Nutrient removal in biological systems. *J. Inst. Water Pollut. Control*, 74, 2–12.

Barnard, J.L. (1976). A review of biological phosphorus removal in the activated sludge process. *Water SA*, 2(3), 135–144.

Barnard, J.L. (1984). Activated primary tanks for phosphate removal. *Water SA*, 10(3), 121.

Barnard, J.L. (1985). The role of full-scale research in biological phosphate removal. In *Proceedings of the University of British Columbia Conference on New Directions and Research in Waste Treatment and Residuals Management*, Vancouver, 23–28 June.

Barnard, J.L., Dunlap, P., Steichen, M. (2017). Rethinking the mechanisms of biological phosphorus removal. *Water Environ. Res.*, 89(11), 2032–2043.

Barthélémy, K. (2012). Adsorption des anions phosphate par des composés ferriques en vue du traitement des eaux usées : approche en réacteur homogène et en mode hydrodynamique contrôlé. PhD Thesis, University of Lorraine.

Barthélémy, K., Naille, S., Despas, C., Ruby, C., Mallet, M. (2012). Carbonated ferric green rust as a new material for efficient phosphate removal. *J. Colloid Interface Sci.*, 384, 121–127.

Battistoni, P., Pavan, P. Prisciandaro, M., Cecchi, F. (2000). Cristallization of struvite: A simple method to fix phosphates on the anaerobie supernatant. *Aqua*, 34(11), 3033–3041.

Bellier, N., Chazarenc, F., Comeau, Y. (2006). Phosphorus removal from wastewater by mineral apatite. *Water Res.*, 40(15), 2965–2971.

Besnault, S., Seranne, L., Prost-Boucle, S. (2023). Déphosphatation physicochimique dans les stations d'épuration des petites collectivités – Retours d'expérience et recommandations. Report, INRAE, UR REVERSAAL.

Boelee, N., Temmink, H., Janssen, M., Buisman, C., Wijffels, R. (2011). Nitrogen and phosphorus removal from municipal wastewater effluent using microalgal biofilms. *Water Res.*, 45, 5925–5933.

Boer, B., Hirsch, P., Johns, F., Saul, B., Scurrah, N. (2015). *The Mekong: A Socio-Legal Approach to River Basin Development*. Routledge, London.

Bouamra, F. (2018). Déphosphatation des eaux usées par de nouveaux procédés. PhD Thesis, Ecole Nationale Polytechnique, Algiers.

Bouaoun, D. and Aoun, J. (2004). Élimination du phosphore, de l'ammonium et des nitrates par alternance de phases dans des réacteurs biologiques. *Environ. Ing. Dév.*, 35, 28–32.

Bowden, L.I., Jarvis, A.P., Younger, P.L., Johnson, K.L. (2009). Phosphorus removal from wastewaters using basic oxygen steel slag. *Environ. Sci. Technol.*, 43(7), 2476–2481.

Bowes, M.J., Jarvie, H.P., Halliday, S.J., Skeffington, R.A., Wade, A.J., Loewenthal, M. (2015). Characterising phosphorus and nitrate inputs to a rural river using high-frequency concentration-flow relationships. *Sci. Total Environ.*, 511, 608–620.

Bunce, J.T., Ndam, E., Ofiteru, I.D., Moore, A., Graham D.W. (2018). A review of phosphorus removal technologies and their applicability to small-scale domestic wastewater treatment systems. *Front. Environ. Sci.*, 6, 8–25.

Cabanes, F. (2006). Déphosphatation des effluents : précipitation et valorisation du phosphore. PhD Thesis, Toulouse INPT.

Canler, J.P. and Perret, J.M. (2007). Les clari-floculateurs plus particulièrement utilisés en traitement tertiaire. Technical Document 35, FNDAE, Ministère de l'ecologie, du développement et de l'aménagement durables, CEMAGREF Editions, Paris.

Chen, Y., Peng, C., Wang, J., Ye, L., Zhang, L., Aeng, Y. (2011). Effect of nitrate recycling ratio on simultaneous biological nutrient removal in a novel anaerobic/anoxic/oxic (A^2/O) –biological aerated filter (BAF) system. *Bioresour. Technol.*, 102, 5722–5727.

Comeau, Y. (1990). La déphosphatation biologique, métabolisme microbien. *Sciences et Techniques de l'eau*, 23(1), 47–60.

Comeau, Y. (1997). Procédés et conception de filières de déphosphatation biologique. In *Journées internationales d'étude des eaux µ la déphosphatation des eaux usées*. Cebedoc, Villeurbanne.

Comeau, Y., Hall, K., Hancock, R., Oldham, W. (1986). Biochemical model for enhanced biological phosphorus removal. *Water Res.*, 20, 1511–1521.

Comeau, Y., Bellier, N., Chazarenc, F. (2006). Phosphorus removal from wastewater by mineral apatite, *Water Research*, 40(15), 2965–2971.

Cornel, P. and Schaum, C. (2009). Phosphorus recovery from wastewater: Needs, technologies and costs. *Water Sci. Technol.*, 59, 1069–1076.

Crittenden, J., Trussell, R.R., Hand, D., Howe, K., Tchobanoglous, G. (2005). *MWH's Water Treatment: Principles and Design*. John Wiley & Sons, Hoboken, NJ.

Cydzik-Kwiatkowska, A. and Zielinska, M. (2016). Bacterial communities in full scale wastewater treatment systems. *World J. Microbiol. Biotechnol.*, 32, 66–78.

Davelaar, D. (1993). Ecological significance of bacterial polyphosphate metabolism in sediments. *Hydrobiologia*, 253(1–3), 179–192.

De-Bashan, L.E., Moreno, M., Hernandez, J.P., Bashan, Y. (2002). Removal of ammonium and phosphorus ions from synthetic wastewater by the microalgae *Chlorella vulgaris* co-immobilized in alginate beads with the microalgae growth promoting bacterium *Azospirillum brasilense*. *Water Res.*, 36, 2941–2948.

De-Bashan, L.E., Hernandez, J.-P., Morey, T., Bashan, Y. (2004). Microalgae growth-promoting bacteria as "helpers" for microalgae: A novel approach for removing ammonium and phosphorus from municipal wastewater. *Water Res.*, 38, 466–474.

De Kreuk, M.K., Heijnen, J.J., Van Loosdrecht, M.C.M. (2005). Simultaneous COD, nitrogen, and phosphate removal by aerobic granular sludge. *Biotechnol. Bioeng.*, 90, 761–769.

Dernat, M., ElMerich, P., Pouillot, M. (1994). Vers une optimisation de la déphosphatation physico-chimique. In *Actes de la conférence sur la déphosphatation*, Paris, 1–14.

Deronzier, G. and Choubert, J.-M. (2002). Traitement du phosphore dans les petites stations d'épuration à boues activées. Report, FNDAE, Ministère de l'agriculture, de l'alimentation, de la pêche et des affaires rurales.

Deronzier, G. and Choubert, J.-M. (2004). Traitement du phosphore dans les petites stations d'épuration à boues activées. Technical Document 29, FNDAE, Ministère de l'agriculture, de l'alimentation, de la pêche et des affaires rurales, CEMAGREF Editions, Paris.

Díez-Montero, R., De Florio, L., González-Viar, M., Herrero, M., Tejero, I. (2016). Performance evaluation of a novel anaerobic-anoxic sludge blanket reactor for biological nutrient removal treating municipal wastewater. *Bioresour. Technol.*, 209, 195–204.

Drizo, A., Frost, C.A., Grace, J., Smith, K.A. (1999). Physico-chemical screening of phosphate removing substrates for use in constructed wetland systems. *Water Res.*, 33, 3595–3602.

Durrant, A.E., Scrimshaw, M.D., Stratful, I., Lester, J.N. (1999). Review of the feasibility of recovering phosphate from wastewater for use as a raw material by the phosphate industry. *Environ. Tech.*, 20, 749–758.

Ekama, G.A., Siebritz, I.P., Marais, G.v.R. (1983). Considerations in the process design of nutrient removal activated sludge processes. *Wat. Sci. Tech.*, 15, 283–318.

European Council (2000). Directive 2000/60/EC of the European Parliament and of the Council of 23 October 2000 establishing a framework for community action in the field of water policy. *Off. J. Eur. Parliam.*, L327, 1–82.

Fu, Z., Yang, F., An, Y., Xue, Y. (2009). Simultaneous nitrification and denitrification coupled with phosphorus removal in a modified anoxic/oxic-membrane bioreactor (A/O-MBR). *Biochem. Eng. J.*, 43, 191–196.

Gaid, K. (2025a). *Wastewater Reuse 1: Characteristics, Uses, Applications, Filtration and Disinfection of Water*. ISTE Ltd, London and Wiley, Hoboken, NJ.

Gaid, K. (2025b). *Wastewater Reuse 2: Micropollutants, Membranes and Treatment Procedures*. ISTE Ltd, London and Wiley, Hoboken, NJ.

Garcia Martin, H., Ivanova, N., Kunin, V., Warnecke, F., Barry, K.W., McHardy, A.C. (2006). Metagenomic analysis of two enhanced biological phosphorus removal (EBPR) sludge communities. *Nat. Biotechnol.*, 24, 1263–1269.

Garzon-Zuniga, M.A. and Gonzalez-Martinez, S. (1996). Biological phosphate and nitrogen removal in a biofilm sequencing batch reactor. *Water Sci. Technol.*, 34, 293–301.

Ge, Z., Wei, D., Zhang, J., Hu, J., Liu, Z., Li, R. (2019). Natural pyrite to enhance simultaneous long-term nitrogen and phosphorus removal in constructed wetland: Three years of pilot study. *Water Res.*, 148, 153–161.

Gieseke, A., Arnz, P., Amann, R., Schramm, A. (2002). Simultaneous P and N removal in a sequencing batch biofilm reactor: Insights from reactor- and microscale investigations. *Water Res.*, 36, 501–509.

Gonzalez, L.E., Caizares, R.O., Baena, S. (1997). Efficiency of ammonia and phosphorus removal from a Colombian agro-industrial wastewater by the microalgae *Chlorella vulgaris* and *Scenedesmus dimorphus*. *Bioresour. Technol.*, 60, 259–262.

Gray, S., Kinross, J., Read, P., Marland, A. (2000). The nutrient assimilative capacity of maerl as a substrate in constructed wetland systems for waste treatment. *Water Res.*, 34(8), 2183–2190.

He, S. and Xue, G. (2010). Algal-based immobilization process to treat the effluent from a secondary wastewater treatment plant. *J. Hazard. Mater.*, 178, 895–899.

Huang, W.-Y., Li, D., Yang, J., Liu, Z.-Q., Zhu, Y., Tao, Q., Xu, K., Li, J.Q., Zhang, Y.-M. (2013). One-pot synthesis of Fe (III)-coordinated diamino-functionalized mesoporous silica: Effect of functionalization degrees on structures and phosphate adsorption. *Microporous Mesoporous Mater*, 170, 200–210.

Huang, W.-Y., Li, D., Liu, Z.-Q., Tao, Q., Zhu, Y., Yang, J., Zhang, Y.-M. (2014). Kinetics, isotherm, thermodynamic, and adsorption mechanism studies of La(OH)3-modified exfoliated vermiculites as highly efficient phosphate adsorbents. *Chem. Eng. J.*, 236, 191–201.

Izadi, P., Izadi, P., Eldyasti, A. (2020). Design, operation and technology configurations for enhanced biological phosphorus removal (EBPR) process: A review. *Rev. Environ. Sci. Bio/Technol.*, 19(3), 561–593.

Jabari, P., Munz, G., Oleszkiewicz, J.A. (2014). Selection of denitrifying phosphorous accumulating organisms in IFAS systems: Comparison of nitrite with nitrate as an electron acceptor. *Chemosphere*, 109, 20–27.

Jenkins, D., Ferguson, J.F., Menar, A.B. (1971). Chemical processes for phosphate removal. *Water Res.*, 5(7), 369–389.

Jenkins, D., Yilmaz, S., Keasling, J.D. (2004). PHA synthases from a mixed culture performing enhanced biological phosphorus removal. *Environmental Biotechnology ESEB*, 6, 65–68.

Ji, B., Zhang, M., Wang, L., Wang, S., Liu, Y. (2020). Removal mechanisms of phosphorus in non-aerated microalgal-bacterial granular sludge process. *Bioresour. Technol.*, 312, 123531.

Johansson, L. and Gustafsson, J.P. (2000). Phosphate removal using blast furnace slags and opoka-mechanisms. *Water Res.*, 34(1), 259–265.

Kaasik, A., Vohla, C., Motlep, R., Mander, U., Kirsimae, K. (2008). Hydrated calcareous oilshale ash as potential filter media for phosphorous removal in constructed wetlands. *Water Res.*, 42(4–5) 1315–1323.

Kim, T.-H., Lee, Y., Han, S.-H., Hwang, S.-J. (2013). The effects of wavelength and wavelength mixing ratios on microalgae growth and nitrogen, phosphorus removal using *Scenedesmus* sp. for wastewater treatment. *Bioresour. Technol.*, 130, 75–80.

Kpannieu, D.E. (2018). Traitement de la pollution phosphorée des effluents résiduaires urbains par des matériaux de Côte d'Ivoire constitués de schiste ardoisier, grès et latérite : approches en réacteur homogène, en mode hydrodynamique contrôlé et en réacteur micro-pilote de terrain. Thesis, University of Lorraine, Lorraine and University of Abobo-Adjamé, Abidjan.

Kuba, T., Smolders, G., van Loosdrecht, M.C.M., Heijnen, J.J. (1993). Biological phosphorus removal from wastewater by anaerobic-anoxic sequencing batch reactor. *Water Sci. Technol.*, 27, 241–252.

Larsdotter, K. (2006). Wastewater treatment with microalgae – A literature review. *Vatten*, 62, 31–38.

Lee, S.I., Weon, S.Y., Lee, C.W., Koopman, B. (2003). Removal of nitrogen and phosphate from wastewater by addition of bittern. *Chemosphere*, 51(4), 265–271.

Lemercier, B., Saby, N., Schwartz, C., Walter, C. (2003). Analyse de l'évolution temporelle des teneurs en carbone organique et phosphore extractible des sols français. In *8èmes Journées Nationales d'Etude des Sols*, Angers.

Letterman, R.D. and Yiacoumi, S. (2011). Coagulation et floculation. In *Water Quality and Treatment: A Handbook on Drinking Water*, 6th edition, Edzwald, J.K. (ed.). McGraw Hill, New York.

Letterman, R.D., Amirtharajah, A., O'Melia, C.R. (1999). Coagulation and flocculation. In *Water Quality and Treatment: A Handbook of Community Water Supplies*, 5th edition, Letterman, R.D. (ed.). American Water Works Association and McGraw Hill, New York.

Lewandowski, J., Schauser, I., Hupfer, M. (2003). Long term effects of phosphorus precipitations with alum in hypereutrophic Lake Süsser See (Germany). *Water Research*, 37(13), 3194–3204.

Li, J., Xing, X.H., Wang, B.Z. (2003). Characteristics of phosphorus removal from wastewater by biofilm sequencing batch reactor (SBR). *Biochem. Eng. J.*, 16, 279–285.

Li, J., Li, B. Huang, H., Lv, X., Zhao, N., Guo, G., Zhang, D. (2019). Removal of phosphate from aqueous solution by dolomite-modified biochar derived from urban dewatered sewage sludge. *Sci. Total Environ.*, 687, 460–469.

Liu, W.T., Mino, T., Matsuo, T., Nakamura, K. (1996). Biological phosphorus removal processes effect of pH on anaerobic substrate metabolism. *Wat. Sci. Tech.*, 34(1–2), 25–32.

Lutterbeck, C.A., Kist, L.T., Lopez, D.R., Zerwes, F.V., Machado, Ê.L. (2017). Life cycle assessment of integrated wastewater treatment systems with constructed wetlands in rural areas. *J. Clean. Prod.*, 148, 527–536.

Mallet, M., Barthélémy, K., Ruby, C., Renard, A., Naille, S. (2013). Investigation of phosphate adsorption onto ferrihydrite by X-ray photoelectron spectroscopy. *J. Colloid Interface Sci.*, 407, 95–101.

Martin, B.D., Parsons, S.A., Jefferson, B. (2009). Removal and recovery of phosphate from municipal wastewaters using a polymeric anion exchanger bound with hydrated ferric oxide nanoparticles. *Water Sci. Technol.*, 60, 2637.

Maurer, M. and Boller, M. (1998). Modelling of phosphorus precipitation in wastewater treatment plants with enhanced biological phosphorus removal. In *Fourth Kollekolle Seminar on Activated Sludge Modelling*, Copenhagen, 16–18 March.

May, L., Place, C., O'Malley, M., Spears, B. (2015). The impact of phosphorus inputs from small discharges on designated freshwater sites. Report 170, Natural England, Worcester.

Metcalf & Eddy (2014). *AECOM, Wastewater Engineering: Treatment and Resources Recovery*. McGraw Hill Education, New York.

Molinos-Senante, M., Gómez, T., Garrido-Baserba, M., Caballero, R., Sala-Garrido, R. (2014). Assessing the sustainability of small wastewater treatment systems: A composite indicator approach. *Sci. Total Environ.*, 497, 607–617.

Molle, P., Liénard, A., Boutin, C., Merlin, G., Iwema, A. (2005). How to treat raw sewage with constructed wetlands: An overview of the French systems. *Water Sci. Technol.*, 51(9), 11–21.

Molle, P., Harouiya, N., Prost-Boucle, S., Morlay, C., Esser, D., Martin, S., Besnault, S. (2012). Déphosphatation des eaux usées par filtres plantés garnis de phosphorites. Report, INRAE.

Monclus, H., Sipma, J., Ferrero, G., Rodriguez-Roda, I., Comas, J. (2010). Biological nutrient removal in an MBR treating municipal wastewater with special focus on biological phosphorus removal. *Bioresour. Technol.*, 101, 3984–3991.

Nemery, J. (2017). Phosphore et eutrophisation. *Encyclopédie de l'Environnement.*

Nemery, J. and Garnier, J.A. (2014). Dynamique du phosphore dans le bassin de la Seine et son estuaire. *Tech. Sci. Méthodes*, 4(4), 33–45.

Nesbitt, J. (1969). Phosphorus removal: The state of the art. *Water Pollut. Control Fed.*, 41, 701–713.

Ng, W.J., Ong, S.L., Gomez, M.J., Hu, J.Y., Fan, X.J. (2000). Study on a sequencing batch membrane bioreactor for wastewater treatment. *Water Sci. Technol.*, 41, 227–234.

Nguyen, T., Ngo, H., Guo, W., Zhou, J., Wang, J., Liang, H., Li, G. (2014). Phosphorus elimination from aqueous solution using "zirconium loaded okara" as a biosorbent. *Bioresour. Technol.*, 170, 30–37.

Parsons, S.A. and Smith, J.A. (2008). Phosphorus removal and recovery from municipal wastewaters. *Elements*, 4, 109–112.

Pastorelli, G., Canziani, R., Pedrazzi, L., Rozzi, A. (1999). Phosphorus and nitrogen removal in moving-bed sequencing batch biofilm reactors. *Water Sci. Technol.*, 40, 169–176.

Patel, A., Barrington, S., Lefsrud, M. (2012). Microalgae for phosphorus removal and biomass production: A six species screen for dual-purpose organisms. *Gcb Bioenergy*, 4, 485–495.

Pernitsky, D. and Edzwald, J. (2006). Sélection des coagulants à base d'alun et de polyaluminium : principes et applications. *Aqua*, 55(2), 121–141.

Perret, J.M. and Canler, J.P. (2013). Affinage du traitement de la pollution particulaire par les procédés mécaniques "rustiques". Technical Document 40, FNDAE, IRSTEA, Paris.

Pinay, G., Gascuel, C., Ménesguen, A., Souchon, Y., Le Moal, M., Levain, A., Etrillard, C., Moatar, F., Pannard, A., Souchu, P. (2017). L'eutrophisation : manifestations, causes, conséquences et prédictibilité. Report, Ifremer.

Pronk, M., de Kreuk, M.K., de Bruin, B., Kamminga, P., Kleerebezem, R., van Loosdrecht, M.C.M. (2015). Full scale performance of the aerobic granular sludge process for sewage treatment. *Water Res.*, 84, 207–217.

Quintana, M., Colmenarejo, M.F., Barrera Loredo, J., Garcia, G., Garcia, E., Bastos, A. (2004). Use of a byproduct of magnesium oxide production to precipitate phosphorus and nitrogen as struvite from wastewater treatment liquors. *Agric. Food Chem.*, 52(2), 294–299.

Quintana, M., Colmenarejo, M.F., Barrera Loredo, Hernandez, F.P.S. (2008). Removal of phosphorus through struvite precipitation using a byproduct of magnesium oxide production (BMP): Effect of the mode of BMP preparation. *Chem. Eng. J.*, 136(2–3), 204–209.

Renman, A. and Renman, G. (2010). Long-term phosphate removal by the calcium-silicate material Polonite in wastewater filtration systems. *Chemosphere*, 79, 659–664.

Seo, Y.I., Hong, K.H., Kim, S.H., Chang, D., Lee, K.H., Kim, Y. (2013). Phosphorus removal from wastewater by ionic exchange using a surface-modified Al alloy filter. *J. Ind. Eng. Chem.*, 19, 744–747.

Shi, J., Podola, B., Melkonian, M. (2007). Removal of nitrogen and phosphorus from wastewater using microalgae immobilized on twin layers: An experimental study. *J. Appl. Phycol.*, 19, 417–423.

Simpkins, M.J. and McLaren, A.R. (1978). Consistent biological phosphate and nitrate removal in an activated sludge plant. In *Proceedings of the Ninth International Conference on Water Pollution Research*, Stockholm, 433–442.

Smolders, G.J.F., van der Meij, J., van Loosdrecht, M.C.M., Heijnen, J.J. (1994). Model of the anaerobic metabolism of the biological phosphorus removal process: Stoichiometry and pH influence. *Biotechnol. Bioeng.*, 43, 461–470.

Smolders, G.J.F., van Loosdrecht, M.C.M., Heijnen, J.J. (1995). A metabolic model for the biological phosphorus removal process. *Water Sci. Technol.*, 31, 79–93.

Stricker, A.E. and Heduit, A. (2010). Phosphore des eaux usées : état des lieux et perspectives. Technical Report, Irstea, Paris.

Stumm, W. and Morgan, J.J. (1996). *Aquatic Chemistry*. John Wiley & Sons, Hoboken, NJ.

Sun, L., Wang, Z., Wei, X., Li, P., Zhang, H., Li, M. (2015). Enhanced biological nitrogen and phosphorus removal using sequencing batch membrane-aerated biofilm reactor. *Chem. Eng. Sci.*, 135, 559–565.

Tang, C.-C., Tian, Y., Liang, H., Zuo, W., Wang, Z.-W., Zhang, J., He, Z.-W. (2018). Enhanced nitrogen and phosphorus removal from domestic wastewater via algae-assisted sequencing batch biofilm reactor. *Bioresour. Technol.*, 250, 185–190.

Tchobanoglous, G., Burton, F.L., Stensel, H.D. (2003). *Wastewater Engineering Treatment and Reuse*. McGraw Hill Higher Education, Boston.

UKTAG (2013). Updated recommendations on phosphorus standards for UKTAG.

Vachon, A., Naturale, G., Tondeux, P., Bourdon, F. (1992). La déphosphatation biologique des eaux usées. *Techniques, Sciences et Méthodes*, 87(4), 212–221.

Van Benschoten, J.E. and Edzwald, J.K. (1990). Chemical aspects of coagulation using aluminum salts – I. Hydrolytic reactions of alum and polyaluminum chloride. *Water Res.*, 24(12), 1519–1526.

Van Haandel, A.C. (1981). Nitrification and denitrification kinetics in the activated sludge process. Thesis, University of Capetown.

Van Haandel, A.C., Ekama, G.A., Marais, G.R. (1981). The activated sludge process: 3 singles sludge denitrification. *Water Res.*, 15(10), 1135–1152.

Villebrun, J.F. (1989). La déphosphatation biologique appliquée à la station d'épuration de Craon. Report, DDAF de la Mayenne.

Vohla, C., Kõiv, M., Bavor, H.J., Chazarenc, F., Mander, Ü. (2011). Filter materials for phosphorus removal from wastewater in treatment wetlands – A review. *Ecol. Eng.*, 37, 70–89.

Vymazal, J. (2007). Removal of nutrients in various types of constructed wetlands. *Sci. Total Environ.*, 380, 48–65.

Wagner, M., Amann, R., Lemmer, H., Manz, W., Schleifer, K.H. (1994). Probing activated sludge with fluorescently labeled rRNA targeted oligonucleotides. *Wat. Sci. Tech.*, 29(7), 15–23.

Wentzel, M.C., Lötter, L.H., Loewenthal, R.E., Marais, G.v.R. (1986). Metabolic behaviour of *Acinetobacter* spp. in enhanced biological phosphorus removal – A biochemical model. *Water SA*, 12(4), 209–224.

Wexler, F. (1997). La déphosphatation physico-chimique : application en batch avec sels de fer, hydroxydes et polychlorures d'aluminium. Typologie du phosphore dans les boues biologiques aérées. Technical Report, FNDAE.

Wood, R. and McAtamney, C. (1996). Constructed wetlands for wastewater treatment: The use of laterite in the bed medium in phosphorus and heavy metal removal. *Hydrobiologia*, 340, 323–331.

Xu, D.F., Xu, J.M., Wu, J.J., Muhammad, A. (2006). Studies on the phosphorus sorption capacity of substrates used in constructed wetland systems. *Chemosphere*, 63, 344–352.

Xu, M., Bernards, M., Hu, Z. (2014). Algae-facilitated chemical phosphorus removal during high-density *Chlorella emersonii* cultivation in a membrane bioreactor. *Bioresour. Technol.*, 153, 383–387.

Xue, S., Wei-Jing, Z., Liang, W., Weixiang, W. (2014). Review on the main microorganisms and their metabolic mechanisms in enhanced biological phosphorus removal (EBPR) systems. *Yingyong Shengtai Xuebao*, 25, 892.

Yang, S., Yang, F., Fu, Z., Wang, T., Lei, R. (2010). Simultaneous nitrogen and phosphorus removal by a novel sequencing batch moving bed membrane bioreactor for wastewater treatment. *J. Hazard. Mater.*, 175, 551–557.

Zhu, D., Chen, Y., Yang, H., Wang, S., Wang, X., Zhang, S., Chen, H. (2020). Synthesis and characterization of magnesium oxide nanoparticle containing biochar composites for efficient phosphorus removal from aqueous solution. *Chemosphere*, 247, 125847.

3

Clarification

3.1. Introduction

The activated sludge unit typically consists of an aeration tank where transfer and metabolization processes take place, and a clarifier where the solids (biomass) are separated from the clarified water. This separation stage plays two essential roles: minimizing the suspended solids (SS) content in the clarified water and retaining the purifying biomass for return to the bioreactor to maintain a biomass concentration suitable for the desired treatment. It also involves accumulating and thickening the sludge mass so that it can be extracted in the smallest possible volumes and then sent to the appropriate sludge treatment facility.

Clarifiers are secondary settling tanks in biological wastewater treatment plants, built for the continuous removal of solids deposited by sedimentation. The sludge settles to the bottom of the tank, where it is collected by a scraping mechanism.

3.1.1. *Principle*

Clarifiers are used in activated sludge wastewater treatment plants to separate clarified water from biomass. The sludge settles to the bottom of the tank and is collected by a scraper. The supernatant, or clarified effluent, is discharged into the environment.

There are three functions of the secondary settling tank:

– Clarification: producing a clarified final effluent whose concentration of suspended solids is generally less than 30 $mg.L^{-1}$; that is, a separation efficiency greater than 98%.

– Thickening: producing a continuous flow of concentrated sludge.

– Storage: storing the additional mass of sludge generated by momentary hydraulic overload.

A recirculation system recycles the sludge in the upstream biological tank, which ensures the maintenance of the suspended solid concentration in the biological reactors. The sludge is also extracted by pumping.

3.1.2. *Types of clarifiers*

3.1.2.1. *Shapes*

A clarifier can be rectangular or circular. Rectangular structures have the advantage of being more compact and easier to install. However, the sludge recycling system is more complex, which increases the cost of these structures. For these economic reasons, clarifiers are most often circular.

3.1.2.2. *Sludge extraction*

The settled sludge from the clarifier can be removed by scraping or suction. Scraping is an economical solution suitable for small structures (less than 20 meters in diameter or width). However, for larger structures, easily returning the settled sludge to the central tank requires a steeply sloped channel bottom, thus a deep structure and complex equipment. Suction removal of the sludge is preferred for larger structures, generally those with a diameter of 25 meters or more. The dimensions, equipment and shape of the clarifier must be determined in such a way as to fulfill both of these functions. The reliability of this step is therefore crucial for the quality of the clarified water, as well as for controlling the biological processes by preventing the "leaching" of the biological reactor (a gradual decrease in biomass concentration, and therefore in reactor activity, due to a growth rate of the purifying species that is lower than the bioreactor specific extraction flow rate).

It is therefore essential to ensure proper sludge settling, which, unfortunately, can sometimes be overlooked by operators due to variability in the input composition or the occurrence of suboptimal reaction conditions (accidental introduction of toxins, temperature drop, insufficient dissolved oxygen content, pH deviation, excessive biomass concentration, etc.), leading to deflocculation or the formation of flocs with low settling capacity (sludge bulking). Consequently, it is not uncommon to observe malfunctions in such systems that are detrimental to the environment or to reuse objectives.

3.2. Settling rate of particles in a liquid

Any particle with a density greater than the surrounding liquid will tend to settle. However, particle size, shape and surface properties will impact the settling rate. Table 3.1 shows that particle size has a significant influence on the settling rate.

Particle diameter (mm)	Order of magnitude	Settling time required
10	Gravel	0.3 s
1	Coarse sand	3 s
0.1	Fine sand	38 s
0.01	Silt	33 min
0.001	Bacteria	55 h
0.0001	Colloidal particles	230 days
0.00001	Colloidal particles	6.3 years
0.000001	Colloidal particles	63 years

Table 3.1. *Settling time as a function of particle diameter*

The theory of solid settling in a liquid obeys Stokes's law. It shows that the settling velocity of a particle is proportional to the difference in densities between the particle and the liquid and to the square of the particle's dimension.

Any phenomenon that increases particle diameter promotes settling. Most of the suspended matter in activated sludge is naturally flocculated and can be concentrated by settling.

3.2.1. *Diluted suspensions*

When the volume concentration of suspended solids is less than 100 $mg.L^{-1}$, the particles are sufficiently separated. Each particle settles freely without interference from adjacent particles (i.e., unimpeded sedimentation).

Each settles according to its physicochemical characteristics.

In a fluid, a freely moving solid particle is subject to three forces:

– Weight: a function of its density (ρp), volume and gravitational acceleration (g).

– Archimedes' principle: a function of the water density (g).

– Friction forces: a function of the square of the relative velocity of the solid with respect to the fluid, the master torque (surface area of the projection, equal to $\pi d^2/4$ for a sphere of diameter (d)) and the drag coefficient of the solid in the fluid (Cx), which varies according to the shape and Reynolds number (Re) of the particle.

When a particle settles in a fluid, it accelerates until the drag force due to its motion equals the particle's weight in the fluid. At this point, the particle will have reached its terminal velocity (vs).

An illustration of the settling of an ideal spherical particle is shown in Figure 3.1: vs is the particle settling rate ($m.s^{-1}$); D is Newton's drag force; P is the submerged weight of the particle; d is the particle diameter (m); Sp is the projected surface of the particle normal to the direction of movement (m^2); Vp is the particle volume (m^3); ρp is the particle density ($kg.m^{-3}$); ρe is the water density ($kg.m^{-3}$); μ is the dynamic viscosity of the fluid ($N.s.m^{-2}$); and C_D is the drag coefficient.

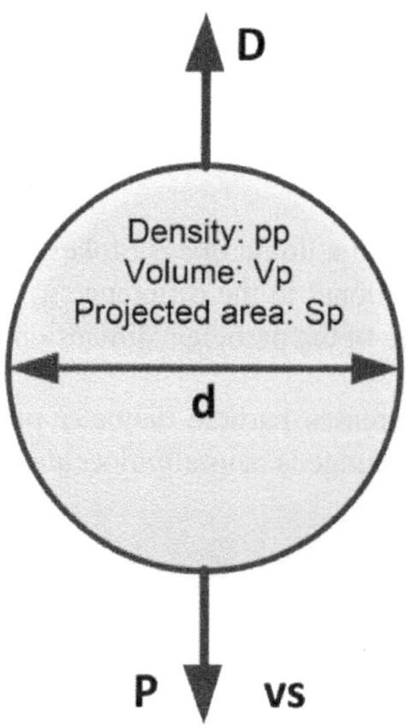

Figure 3.1. *Sedimentation rate of a particle*

The drag force D of the particle is given by:

$$D = \frac{CD}{2} \frac{Sp\, \rho e\, vs^2}{2}$$

The weight of the submerged particle is:

$$P = (\rho p - \rho e)\, g\, Vp$$

Since D = P, when substituting Sp and Vp with their values, we obtain:

$$Vs = \sqrt{\frac{4}{3}\frac{(\rho p - \rho e_l)}{\rho e_l}\frac{gd}{C_D}}$$

C_D is a function of the Reynolds number (Re). For spherical particles, it is given by the following expression:

$$Re < 1,\ C_D = \frac{24}{R_e}$$

$$1 < Re < 10^4,\ C_D = \frac{24}{R_e} + \frac{3}{(R_e)^{1/2}} + 0.34$$

$$10^3 < Re < 10^5\ \ C_D \approx 0.4$$

By replacing the value of Re < 1 in laminar flow, while noting that Re is given by:

$$Re = \frac{d\ \rho e\ vs^2}{\mu}$$

the equation becomes Stokes's equation:

$$vs = \frac{g}{18}\frac{(\rho p - \rho e)}{\mu}d^2$$

For high Re values, where Cd ≈ 0.4, the expression is written as:

$$vs = \sqrt{3.33\frac{(\rho p - \rho e)}{\rho e}gd}$$

In summary, vs depends on a particular diameter, particle density and, under certain conditions, fluid viscosity and therefore temperature, which is important for understanding settling behavior.

Furthermore, in practical settling tanks, the terminal settling velocity is rapidly reached; thus, for non-flocculating particles and uniform fluid flow, the settling velocity is constant throughout the settling time. This fact can be usefully applied to a study of settling in an ideal settling tank, providing an important design principle for settling processes.

The results of numerous experimental studies have led to the categorization of several domains based on the value of Re, as shown in Table 3.2, which also gives the relationships expressing the particle terminal velocity for each domain, as a function of Cx. μ represents the dynamic viscosity of water and d is the particle diameter.

The minimum limit within which Stokes's law remains valid ($Re < 10^{-4}$) corresponds to the distinction made by settling tank operators between a "particle" and a "colloid." In practice, particles smaller than one micrometer have lower limiting settling velocities than $\mu m.s^{-1}$. This speed is of the same order of magnitude as that of Brownian diffusion (thermal agitation).

Re	Cx	vlim	Domain
$< 10^{-4}$	Stokes's law does not apply, because the motion of the particles is influenced by Brownian motion		
10^{-4} to 1	24/Re	$0.545(\rho p - \rho e)\ \mu^{-1} d^2$	Laminar flow
1 to 10	$26/Re^{0.77}$	$0.57\ (\rho p - \rho e)^{0.814}\ \mu^{-0.625}\ d^{1.439}$	Transitional flow
10 to 10^2	$20/Re^{0.65}$	$0.73(\rho p - \rho e)^{0.741}\ \mu^{-0.481}\ d^{1.222}$	Transitional flow
10^2 to 10^3	$4.92/Re^{0.346}$	$1.81(\rho p - \rho e)^{0.604}\ \mu^{-0.209}\ d^{0.813}$	Transitional flow
10^3 to 2 10^5	0.44	$5.40(\rho p - \rho e)^{0.5}\ d^{0.5}$	Turbulent flow

Table 3.2. *Terminal settling velocities of spherical particles depending on the Reynolds number (Blazy et al. 2013)*

3.2.2. *Flocculated particles*

Considering the settling times of isolated particles, the data in Table 3.2, obtained using Stokes's law, show that surface properties become paramount for ultrafine particles. The interaction between particles results in collisions followed by aggregation when repulsive forces, due to surface electric charges, are counterbalanced by attracting van der Waals forces. Destabilizing the suspension of very fine particles therefore amounts to reducing electrostatic repulsive forces. It can be induced by the addition of chemical reagents (coagulants and/or flocculants). Fine particles then form microflocs, which subsequently develop into larger flocs or clumps that settle more easily. The formation of microflocs is governed by Brownian diffusion, while the formation of flocs is governed by the energy dissipated within the volume occupied by a liquid of a given viscosity, thus creating a velocity gradient (Blazy et al. 2013).

It is observed that, for a flocculated suspension, settling occurs in several phases, as is also the case for concentrated suspensions:

– the formation of a well-defined interface;

– a "piston-like" settling of this interface, at a regular speed;

– settling at decreasing speed, called the "sludge compaction phase".

The first phase corresponds to flocculation of the suspension. The second phase is particularly important, as it is within this range of particle concentrations that sludge blanket clarifiers operate. The final phase corresponds to sludge thickening. To facilitate the transition from the first phase to the second phase, it is possible to modify the particle diameter, the density or the flow geometry. The techniques used for this purpose include sedimentation in sludge blanket clarifiers or with sludge recirculation, weighted flocs and lamellar settling.

3.2.3. *Concentrated suspensions*

As soon as the volumetric concentration of solids becomes high (> 0.5%), interactions between particles are no longer negligible. Sedimentation is hindered and the settling rate is slowed. The particles adhere to each other, and the mass settles in a piston-like fashion, with a clear interface between the sludge and the supernatant liquid.

In laminar flow (Re < 1) and in a Newtonian environment, the equivalent kinematic viscosity of a suspension (νm) and its density (ρm) are expressed approximately using those of the liquid (νe and ρe) according to the relations:

$$\nu m = \nu e \frac{D + 1/\rho p}{D - Dlim}$$

$$\rho m = \nu e \frac{D\rho e + 1}{D + (\frac{1}{\rho p})}$$

where D and Dlim are, respectively, the dilutions (volume of liquid per unit mass of solids) of the suspension and the sludge at the end of sedimentation.

By replacing ρe with ρm and νe with νm in Stokes's equation in laminar flow (see Table 3.2), we obtain the approximate value of the terminal settling velocity of particles in concentrated suspension; that is:

$$vlim = 0.545 \frac{(\rho p - \rho e)(D - Dlim)d^2}{((1/D) + \rho e)\left(D + \left(\frac{1}{\rho p}\right)\right)\nu e}$$

The rate of decreasing solid concentration over time is written as:

$$-\frac{dC}{dt} = kC^2$$

which after integration yields:

$$\frac{1}{Co} - \frac{1}{Ct} = kt$$

with Co being the initial solids concentration of the suspension; C the solids concentration at time t; k the flocculation constant; and t the retention time.

3.2.4. *Sedimentation curves*

In activated sludge wastewater treatment plants, the nature of the microorganisms, the floc structure, the effluent viscosity and the particle density play a crucial role in the sedimentation process. Several sludge behavior patterns can be distinguished based on the nature of the particles to be removed and their concentration. Individual particles can be discrete (sand grains) or flocculating (most of the material in biological sludge). Biological sludge concentrations can range from low to high. In the latter case, the particles are in close contact. The flocs that can be sedimented correspond to various physical states.

The schematic diagram shown in Figure 3.2 illustrates the effect of cohesion between particles and the concentration of suspended matter, where the zones of the different classes are defined (Fitch 1979; Ekama et al. 1997). This is a classic sedimentation test, performed with a test tube filled with activated sludge taken from a biological tank. After allowing the suspension to settle, a clear interface forms at the top of the test tube, separating the sludge mass from the clarified supernatant. This interface descends as the suspension settles. At every point, the slope of the sedimentation curve represents the settling rate between the suspension and the clarified supernatant. This leads to the conclusion that when designing clarifiers for activated sludge treatment, surface loading is an important parameter to consider. If it exceeds the floc settling rate, the sludge will be carried into the effluent.

The settling tanks of activated sludge systems are not the same as discrete sedimentation settling tanks, in which the particles are dispersed and the sedimentation mechanisms are entirely different from those found in the settling tanks in sludge recirculation systems.

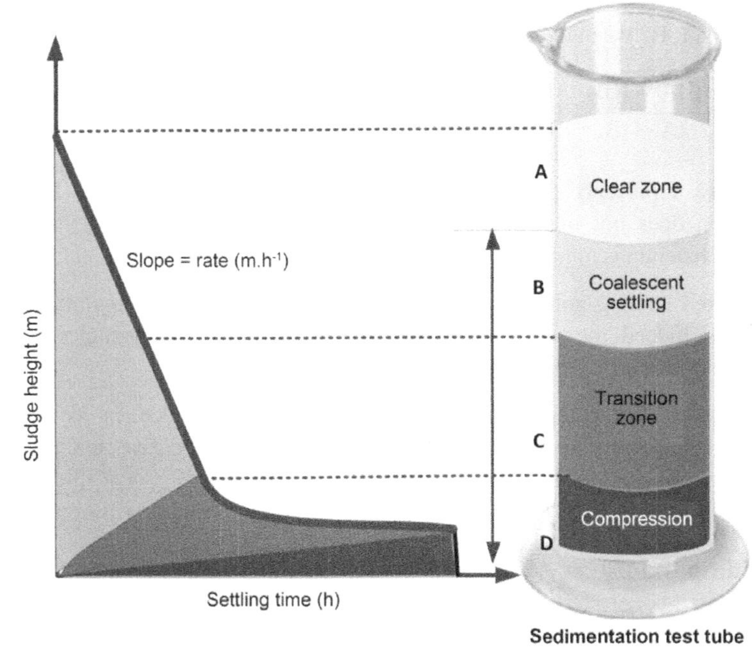

Figure 3.2. *Sedimentation test tube of concentrated sludge*

Figure 3.2 shows all the phenomena observed during a typical laboratory test. Starting with a suspension of uniform particle concentration, a layer of clear liquid (A) appears at the top of the test tube, separated from the suspension by an interface. Simultaneously, a sediment deposit (D) forms at the bottom of the test tube. Between these two zones there is a zone with a dilute concentration gradient (B) and a zone with a concentration gradient (C). The boundary between the last two zones, however, remains somewhat unclear. The curve describes the evolution over time of the position of the interface between the clarified water and the dilute zone (A and B), which should allow us to determine the relationship between sedimentation rate and solid concentrations.

Four distinct zones can be identified between the top and bottom of the test tube, which can be likened to what generally occurs in a clarifier.

The following zones are distinguished:

– Clarified water zone (A): most of the suspended matter has settled, leaving a fairly clear zone at the top of the test tube.

– Coalescing settling zone (B): the flocs begin to gather into flakes and the settling rate is constant. Between the two interfaces A and B, the settling rate is written as:

$$vs = \frac{HA - HB}{tA - tB} = \frac{VA - VB}{S(tA - tB)}$$

This area represents the effects of discrete particle settling. The low solids concentration does not imply significant interactions between particles; that is, the flocculation reaction is minimal, or even negligible.

– Hindered settling zone (C): collisions between flocs lead to agglutination and larger flakes. Indeed, the concentration becomes high enough that particles come into contact as they settle, resulting in a change in their flow pattern.

– Compaction (or compression) zone (D): when the interface, in its downward movement, encounters the upper limit of this zone, its speed decreases and the settling curve becomes asymptotic to a straight line parallel to the x-axis.

3.3. Settling theory

Settling can be defined as the removal of solid particles from a suspension by gravity. Clarification is a similar term, which generally refers specifically to the function of a settling tank in removing suspended solids from the water to produce a clarified effluent. Thickening in clarifiers is the process by which the settled impurities are concentrated and compacted on the bottom of the structure or in the sludge collection hoppers.

The process of gravity settling differs depending on whether the suspension flocculates or not. Materials such as sewage sludge tend to flocculate as they thicken, unlike coarse mineral particles which do not, and the flocculations retain water that can only be expelled from the sediment by compression.

3.3.1. *Fundamental law: Kynch theory*

3.3.1.1. *Relationship between particle height/concentration*

As early as 1952, Kynch showed that the settling process is transient along the height and that there is only one velocity for all solids (Kynch 1952).

A number of researchers (Michaels and Bolger 1962; Shirato et al. 1970; Fitch 1983) noted that Kynch's theory is not valid in the compression zone, where the velocity of the particles depends on the stress gradient of the solids.

Kynch developed an original method that requires only a single batch test. He proposed the following theory: the settling velocity of rigid particles in free settling depends only on the local solid concentration of the suspension. It is a decreasing function of the solid concentration. Particles of the same size, shape and density settle at the same rate, and their concentration is uniform for each level in the sedimentation column.

Thus, based on test tube experiments, Kynch plotted on the (H, t) diagram the isoconcentration lines corresponding to the levels where the concentration had a given value X, and their displacement over time (see Figure 3.3). Consider any thin layer X in this area with a suspension concentration X. Such a layer is bounded on its upper surface by a layer of lower concentration and on its lower surface by a layer of higher concentration. The position of this layer changes over time. Since the settling rate of a particle layer depends on its local concentration and is not influenced by neighboring concentrations, there is only one settling rate (v) for each solid concentration (X).

We represent the isoconcentration lines for which the rates are equal to the slopes as follows:

$$dh/dt = v(X)$$

with v(X) being the speed of a plane of constant concentration (C), relative to the test tube.

In the AOB triangle, the settling rates are constant. The concentration is uniformly equal to Xo, corresponding to the beginning of the curve h(t). The equation can be written as follows, for $X_A < X < X_B$:

$$h = ho + v(X)t$$

The isoconcentration lines located in this triangle represent the progressive increase in the concentration of the suspension at speed v(X):

$$v(X) = -dh/dt$$

with X being the local sludge concentration in the area considered. This initial settling rate is therefore given by the slope of the line between points A and B.

Then, the particles accumulate at the bottom of the column, forming superimposed layers of increasing concentration. Leaving aside the linear portion of the curve, Kynch focused his attention on the section of the curve between points B and C. A gradual slowing of the settling rate is observed, and the isoconcentration lines form a bundle of straight lines passing through the origin in triangle BOC. Velocities decrease due to collisions between particles, forming triangle BOC.

The conclusions to his theory are surprisingly simple. By drawing the tangent to the curve at the inflection point (C), its slope represents the settling rate, while the intersection with the y-axis allows us to deduce the local solid concentration. The settling rate vs in the zone of hindered settling is a function solely of the gravimetric concentration.

This method represents a significant advance in the approach to sizing settling tanks, even though it continues to rely on flux theory, which is considered insufficient when explaining the solid concentration profile at the different levels of a continuous settling tank.

Increasing concentrations of B to C correspond to decreasing settling rates vs = f(X). These settling rates are the successive tangents to the curve BC.

This means that from the start of settling, the layers near the bottom pass through all the concentrations between the initial concentration Xo and that corresponding to the point at which compression begins (Xc). The equation of these lines is written as:

$$H = v(X)t$$

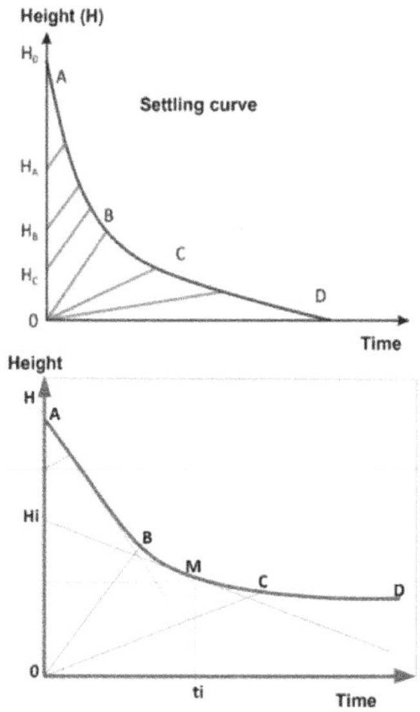

Figure 3.3. *a) Sludge settling and b) determining the concentrations*

The evolution of rates as a function of concentrations can be modeled using an exponential model, the linearization of which provides access to the values of k and n:

$$vs = k.X^{-n}$$

The intersection of the tangent to the settling curve with the x-axis defines a time t, the settling time, while the intersection with the y-axis defines a height H corresponding to the height that the sludge would have if all the particles had the same concentration X as the interface layer.

Within a concentrated suspension, the settling particles are considered individually, with the fastest being slowed down by the slowest. This settling movement is accompanied by a displacement of water of an equivalent volume flowing upwards, thus decreasing the settling rate.

This phenomenon is generally represented by the existence of a clear interface between a clarified supernatant and a fluid phase containing almost all of the suspended solids.

In a cylindrical test tube of volume V and cross-section S, at time t0, the suspension is considered homogeneous at a concentration X0 and all particles are moving at a velocity v0. Between t0 and the final time point (> t_c), the interface separating the clarified phase and the sludge phase moves at a speed that is considered constant and proportional to the concentration at the interface, while the settled sludge (the heaviest) accumulates at the bottom. At infinity $t\infty$, the sludge deposited at the bottom thickens, corresponding to the compaction phase.

At time zero, the suspension has a uniform concentration Xo and an initial height Ho. During a stabilization time t_2, the solid–water interface drops to a height Hi; a concentration over time, Xi ti.

If S is the cross-sectional area of the column, the total mass of solids in the column can be written as:

$$Xo\ Ho\ S = Xi\ vi\ S\ ti$$

If vi = Hi/ti, we can deduce that:

$$Xo\ Ho = Xi\ Hi$$

Thus, from a settling test in a graduated cylinder or column, the settling rate corresponding to a given concentration can be calculated.

3.3.1.2. *Height/concentration*

The settling rate of the interface at each instant is represented by the slope of the tangent to the curve at that point:

$$v = \frac{dh}{dt} = \frac{1}{S}\frac{dV}{dt}$$

Kynch's theory applies to phases two and three. F ($kg.m^{-2}.h^{-1}$) is the flux of sludge particles crossing the unit area during the unit time, and can be written as:

$$F = v\,X$$

The accumulation of solids between two heights h and (H + dh) gives:

$$F\,(H + dh) - F\,(h) = \frac{\partial}{\partial t}(X\,dh)$$

that is:

$$(F\,(H + dh) - F\,(h))\,dt = \frac{\partial X}{\partial t}\,dh\,dt$$

and:

$$\frac{\partial F}{\partial h} = \frac{\partial X}{\partial t} = \frac{\partial F}{\partial X}\cdot\frac{\partial x}{\partial h}$$

Since F is a function of X alone, this becomes:

$$\frac{\partial x}{\partial h} + \frac{v(X)\,\partial xX}{\partial h} = 0 \text{ if} - v(X) = \partial F/\,\partial X$$

Since:

$$dx = \frac{\partial X}{\partial h}dh + \frac{\partial X}{\partial t}dt = 0$$

it becomes:

$$\frac{\partial X}{\partial t} + \frac{dh}{dt}\frac{\partial X}{\partial h} = 0$$

Thus, v(X) = dh/dt, which shows that the straight line equation with constant sludge concentration passing through the origin is:

$$H = v(X)t$$

The total sludge that has passed through the layer at constant concentration along with that initially present is written as:

$$\frac{hoxo}{x} = vt + Vt = h - \frac{dh}{dt}t$$

which translates to HoXo = H X, obtained on a curve H(t). A settling curve obtained in the laboratory allows us to calculate v(X).

3.3.1.3. *Geometric determination of the unit area*

Consider a settling curve of the sludge representing H as a function of t (see Figure 3.4).

We can write:

$$Ho\ Xa = H\ X = Hb\ Xb$$

where Ho and Hb are the relative settling heights at the sludge inlet and outlet, respectively, and Xa and Xb are the sludge concentrations at the inlet and outlet, respectively.

Let C be the point of intersection of the tangent to the settling curve. It corresponds to time tb. The settling rate is then:

$$vs = \frac{H - Hb}{tb}$$

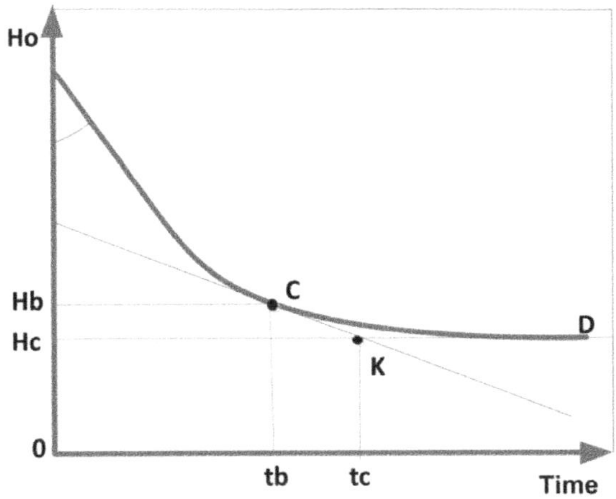

Figure 3.4. *Geometric determination of the unit area*

Substituting vc with the value in the unit area equation, we get:

$$\frac{S}{Qe\ Xa} = \frac{tb}{Ho - Hb}\left(\frac{1}{Xa} - \frac{1}{Xb}\right)$$

If Ho Xa = H X = HbXb, the equation is written as:

$$\frac{S}{Qe\ Xa} = \frac{tb}{\frac{HoXa}{Xa} - \frac{HoXa}{Xb}}\left(\frac{1}{Xa} - \frac{1}{Xb}\right)$$

So:

$$\frac{S}{Qe\ Xa} = \frac{tb}{HoXa\left(\frac{1}{Xa} - \frac{1}{Xb}\right)}\left(\frac{1}{Xa} - \frac{1}{Xb}\right)$$

$$\frac{S}{Qe\ Xa} = \frac{tb}{HoXa} = \frac{tb}{HbXb}$$

The flow rate of purified water Qs is equal to:

$$Qs = Qe - Qb$$

Qe is the input flow rate; Qb is the flow rate of extracted sludge.

Furthermore:

$$Qe\ Xa = Qb\ Xb \text{ and } Hb\ Xb = Ho\ Xa$$

We deduce:

$$Qe = Qs + Qb$$

$$Qe = Qs + \frac{QeXa}{Xb} = \frac{\frac{Qs}{Xb - Xa}}{Xb}$$

$$Qe = \frac{QsXb}{Xb - Xa}$$

Also:

$$Qe = \frac{QsHo}{Ho - Hb}$$

which is reported in the unit area equation:

$$\frac{S}{Qe\,Xa} = \frac{S}{Xa\,\frac{QsHo}{Ho - Hb}}$$

which becomes:

$$\frac{S}{Qs} = \frac{tb}{Ho - Hb}$$

This relationship highlights the importance of surface loading in the calculation of settling tanks:

$$\frac{Qs}{S} = \frac{Ho-Hb}{t}\,m.h^{-1}$$

3.4. Mathematical interpretation of the different types of settling

3.4.1. *Free or coalescent settling (zone B, Figure 3.2)*

Coalescing settling refers to the settling of flocculated particles, which have varying sizes and settling rates, due in particular to flocculation. It concerns dilute solutions whose particles tend to agglomerate and flocculate together, thus increasing the settling rate.

The flakes formed during flocculation behave like particles larger than the individual particles and with a density intermediate between that of a solid and a liquid. The particle density is between 1 and 1.2. The agglomeration of the particles promotes their settling. The greater the agglomeration:

– the higher the suspension concentration, and the more the particle size and mass differ;

– the lower the viscosity of the suspension fluid.

The particle flocculation rate is a function of:

– the distance that the particles travel;

– the surface loading and velocity gradients in the tank;

– the particle concentration;

– the particle diameter.

The settling rate (vs) depends on the direction of liquid flow (horizontally, vertically, circularly, or spirally), the type of flow (laminar or turbulent), the viscosity of the medium and the force, diameter and specific mass of the particles.

A spherical particle of specific mass ρp falls with increasing velocity until the force of attraction and the force of resistance due to friction are balanced. The settling velocity vc then becomes constant.

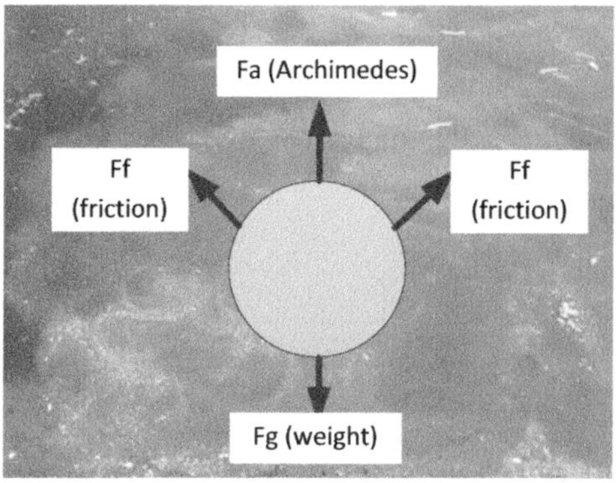

Figure 3.5. *Forces acting on a particle*

The settling velocity of one-dimensional particles in a fluid is considered equivalent to that of an isolated particle as long as the average distance between particles is greater than 10 times their diameter (concentration less than 1 $g.L^{-1}$) and does not depend on the fluid properties or the particle characteristics. The displacement of a particle is described by the applied force F, proportional to the mass m and its acceleration, equal to the resultant of its weight Fg minus the Archimedes thrust Fa and the friction force Ff.

By applying Newton's second law of motion, it is possible to obtain the general equation of motion of a solid particle in a fluid, which is governed by the balance of forces acting on the particle:

$$F = m\, dv/dt$$

$$F = m\, dv/dt = m\gamma = Fg - Fa - Ff$$

When a particle is immersed in a fluid, the particle accelerates until the frictional and buoyancy forces exerted by the fluid equal the gravitational force.

The acceleration of this particle then becomes zero and its settling rate reaches its terminal value:

$$0 = Fg - Fa - Ff$$

1) Gravitational force:

$$Fg = \rho_S \, Vp. \, g$$

with:

- ρs: density of the solid ($kg.m^{-3}$);
- Vp: particle volume (m^3);
- g: acceleration due to gravity (9.81 $m.s^{-2}$).

2) Archimedes:

$$Fa = \rho e \; V.g$$

where ρe is the density of the liquid.

For a spherical particle, the force Fg due to the acceleration field is defined as follows:

$$Fg = \pi \, d^3/6 \, (\rho s - \rho e) \, K.g$$

with K = 1 for the gravitational field.

3) The frictional force opposed by the fluid is a function of the particle size, shape and roughness and the liquid viscosity:

$$Ff = 1/2 \, (C_D \, A \, \rho e.v^2 s)$$

with:

- A: projection surface of particles onto a plane perpendicular to the displacement (m^2);

- C_D: coefficient of friction (drag) depending on the hydrodynamic regime (= b/Re) and the particle shape:

$$m.dv/dt = K.g \, (\rho s - \rho e) \, V - (C_D \, A \, \rho e.v^2 s)/2$$

During gravitational settling, the particle reaches its final settling rate when the accelerating force is reduced to zero:

$$m\,(dv/dt) = 0 = g\,V\,(\rho s - \rho e) - (Cd\,A\,\rho e\,v^2 s)/2$$

with K = 1, and the settling speed is as follows:

$$v^2 s = \frac{2g}{CD}\,\frac{\rho p + \rho e}{\rho e}\,\frac{Vp}{A}$$

where vs is the settling velocity of a particle ($cm.s^{-1}$); g is the acceleration due to gravity (9.81 $m.s^{-2}$); ρs and ρe are the specific masses of the particle and the liquid ($g.cm^{-3}$); Vp is the volume of a particle (cm^3); A is the cross-section of a particle (cm^2); and finally CD is a resistance coefficient, also called Newton's drag coefficient.

At a constant speed, the expression becomes:

$$vs = \sqrt{\frac{2g(\rho s - \rho e)V}{CD\,\rho e\,A}}$$

If we consider spherical particles of diameter d, volume $V = \pi d^3/6$ and projected surface $A = \pi d^2/4$, then V/A = 2d/3.

In turbulent flow, the equation becomes:

$$vs = \left(\frac{4dg}{3}\,\frac{\rho p - \rho e}{\rho e}\,\frac{1}{CD}\right)^{1/2} \quad \text{(Newton's law)}$$

The drag coefficient CD is a function of the hydrodynamic regime, characterized by the Reynolds number (Re) of the particle:

$$Re = vs\,d/\gamma$$

with μ being the dynamic viscosity (Pa.s) and γ the kinematic viscosity of the fluid:

$$\gamma = \mu/\rho e$$

– If $Re < 10^{-4}$ (domain in which the motion of particles is influenced by Brownian motion), the velocity law is not applicable.

In the case we are considering, the Reynolds number is less than 1, and in this case, the drag coefficient can be approximated by CD ≈ 24/Re. Indeed, if $10^{-4} < Re < 1$ is the domain of laminar flow, CD = 24/Re.

Newton's law becomes Stokes's law. The terminal settling velocity, v, is given by Stokes's law:

$$vs = \frac{d^2\ (\rho s - \rho e)g}{18\ v\ \rho e} = \frac{d^2\ (\rho s - \rho e)g}{18\ \mu}$$

The falling speed is proportional to the square of the particle diameter. The particle diameter is less than 10^{-4} m. For K = 1 and g = 9.81 m.s^{-2}:

$$vs = 0.545\frac{d^2\ (\rho s - \rho e)}{v\ \rho e} = 0.545\frac{d^2\ (\rho s - \rho e)}{\mu}$$

– If $0.5 < Re < 10^3$, the flow is transitional, and CD is complex and varies continuously. The particle diameter is between 10^{-4} m and 10^{-3} m.

The following empirical expression (Fair et al. 1968) is proposed:

$$CD = \frac{24}{Re} + \frac{3}{\sqrt{Re}} + 0.34 = \frac{18.5}{Re^{0.6}}$$

therefore:

$$vc = gd^{2\frac{(rs-re)}{18\ re}} + \frac{1}{1 + 0.18\ Re^{0.687}} = [2.32\ (\rho s - \rho e)d^{1.6}\ \rho^{-0.4}\ \mu^{-0.6}]^{0.714}$$

If $10^3 < Re < 2.10^5$, we have the turbulent regime domain, and CD is independent of the Reynolds number. CD = 0.4; hence:

$$vc = \sqrt{\frac{3.3g\ d\ (\rho s - \rho e)}{\rho e}}$$

The settling velocity is proportional to the square root of the diameter. The particles have a diameter > 1 mm.

– If $Re > 2.10^5$, the speed law is no longer applicable.

For all non-laminar flow conditions, the drag coefficient also depends on the particle shape and must be determined experimentally. Non-spherical particles will settle more slowly than spherical particles of the same volume and density.

The relationship between the drag coefficient and the settling velocity vs is summarized in Figure 3.6 using the Reynolds number. It shows the settling velocity of spherical particles in water at 10°C.

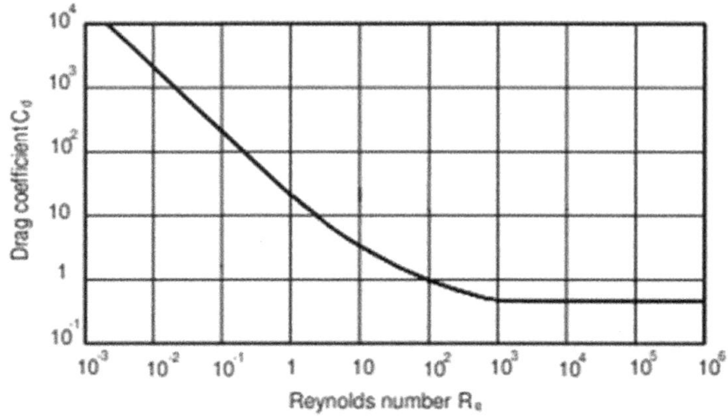

Figure 3.6. *Drag coefficient CD as a function of the Reynolds number Re*

EXAMPLE.–

We want to remove all particles with a diameter of 0.08 mm from wastewater. The water flow rate is 720 $m^3.h^{-1}$, and the specific mass of suspended matter is 1,200 $kg.m^{-3}$. In Figure 3.7, we can easily read a settling velocity vs = 0.05 $cm.s^{-1}$. Thus, vs = 1.8 $m.h^{-1}$.

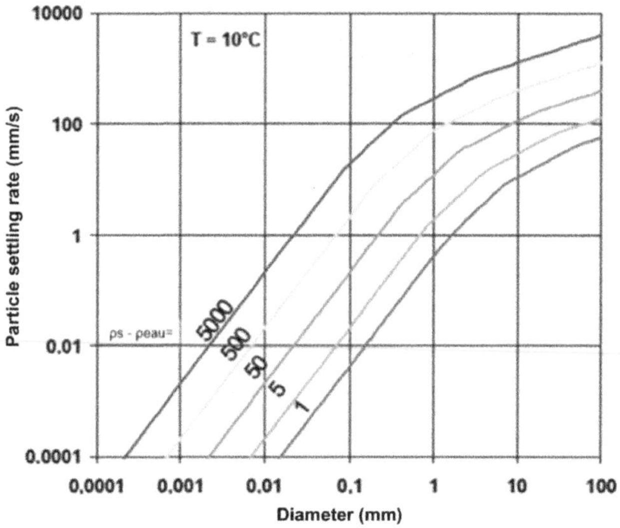

Figure 3.7. *Settling velocity of spherical particles in water (at 10°C)*

Calculating using Stokes's law gives the Poiseuille viscosity 10^{-3}:

$$vs = \frac{9.81}{18} \frac{1\,200 - 1\,000}{0.001} (0.08\ 10^{-3})^2 = 6.7\ 10^{-4}\ \text{m.s}^{-1} = 25\ \text{m.h}^{-1}$$

3.4.2. *Hindered settling (zone C, Figure 3.2)*

This type of settling applies to concentrated solutions such as the sludge leaving the aeration tank and heading towards the secondary settling tank. This solution must go from a sludge concentration of several $g.L^{-1}$ to a maximum value of 30 $mg.L^{-1}$.

Hindered settling is the term used to describe the sedimentation behavior of highly flocculating suspensions such as activated sludge. When the concentration of these suspensions exceeds approximately 1,000 $mg.L^{-1}$, the flocs form a compact mass, creating a distinct interface between the settled zone and the supernatant water.

The clarifier's sizing is based on flow theory. A continuous settling tank (see Figure 3.8) is supplied by a volumetric flow rate Qe with a solid concentration Xa. The extracted flow rates are, respectively:

– sludge flow rate Qb at concentration Xb;

– flow rate Qf, which is released from the tank with a very low suspended solid concentration.

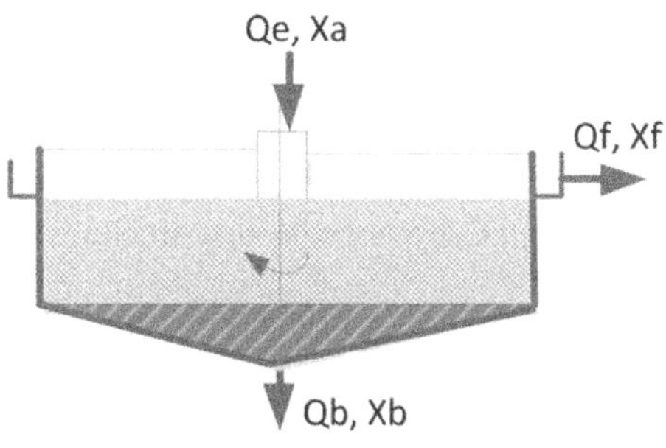

Figure 3.8. *Hydraulic balance on a cylindrical clarifier*

In steady state, the material balance of the clarifier is written as:

Qe Xa = Qb Xb

At the intermediate concentration level Xi (to which corresponds a settling velocity vi), the downward flow of solids G has two components:

– a flow due to the subsidence flow rate, equal to Qb Xi/S;

– a flow due to settling itself, equal to:

$$G = \frac{Qb\ Xi}{S} + vi\ Xi$$

Regarding sludge extraction, we can write:

$$G = \frac{Qb\ Xi}{S}$$

By removing Qb from these two equations, we obtain:

$$G = vi / \frac{1}{Xi} - \frac{1}{Xb}$$

If the settling velocity vi is solely a function of the concentration Xi, then it takes the same value in a batch laboratory test as in a continuous operation. Therefore, settling velocities can be determined by various concentrations. For a fixed Cb value, the flow values G at the various concentrations considered can be calculated. The flow curve (G as a function of X) reaches a minimum at a certain critical concentration Xc. The maximum solid flow that can pass through the clarifier to the extraction point is Gc, and the critical flow corresponding precisely to this concentration.

If the solid supply (flow equal to Q Xa/S) is greater than this critical flux, a critical concentration zone Xc forms in the clarifier: The solids cannot pass through this critical zone at a speed as high as that at which they enter. They therefore accumulate in the critical zone until it eventually overflows.

To prevent this overflow of solids, it is essential to equip the clarifier with a sufficient surface area so that, at any point in the structure, the flow rate does not exceed the critical flow rate:

$$\frac{Qb\ Xa}{S} < vi / \frac{1}{Xi} - \frac{1}{Xb}$$

It is therefore necessary to find the minimum of the expression on the right, a minimum which corresponds to the critical flow Gc; this constitutes the maximum solid load that the clarifier can withstand.

3.4.2.1. *Coe and Clevenger method*

Although the method is old, the proposed relationship is commonly used to determine the surface area required to ensure the settling and thickening of a suspended solids mass. For the operation to be successful, the overflow from the settling tank must consist of clarified effluent. Consequently, if the clarification zone is too shallow, fine particles can be carried into the overflow (Coe and Clevenger 1916).

In a continuous and steady-state operation, the flow rate of the effluent at the outlet (Qf) is equal to the difference between the sludge flow rate at the clarifier inlet (Qe) and the flow rate of sludge extracted from the bottom of the structure (Qb).

If D is the liquid/solid mass ratio, or dilution, at a given point in the clarifier-thickener, DB is the dilution of the underflow and Qe is the mass flow rate of entering dry solids, the mass flow rate of the effluent exiting (Qf) is:

$$Qf = (D - DB)\ Qb$$

and the upward velocity of the water (vup) is written as:

$$vup = \frac{(D - DB)Qb}{S\ \rho e}$$

with S being the surface area of the clarifier-thickener and ρe the density of the effluent.

At equilibrium, the upwelling speed must always be less than the settling velocity.

Considering solid flows (Jernqvist 1965, 1966; Rivet 1981), where X is the solid concentration, vs the relative settling velocity of the sludge and vb the settling velocity of the sludge following extraction, the treated solid flow G results from both the settling solids and their downward movement:

$$G = Xb\ (vb + vs)$$

since:

$$vs = Qb/S$$

and:

$$G = Xb\ (vb + Qb/S)$$

At equilibrium, in steady state, there is no accumulation of solids in the unit. We can write:

$$G = Xb\ Qb/S$$

By combining the two equations, we get:

$$G = \frac{vs}{\frac{1}{Xa} - \frac{1}{Xb}}$$

The solid flow G that can be treated in a surface clarifier-thickener S is therefore a function of the concentration of the extracted sludge (Xb) and that of the entering sludge (Xa) and the settling velocity. But since vs is a function of Xa and the supplied sludge concentration Xa is lower than the concentration of the extracted sludge (Xb), Coe and Clevenger determine vs for a number of values between Xa and Xb and calculate G for these values.

G passes through a minimum (Gmin) which allows access to the surface of the clarifier-thickener:

$$S = Qa/Gmin$$

This finding is consistent with the theory that was later developed by Kynch.

Fitch (1995) simplifies the Coe and Clevenger method. The compression point C is determined on the settling curve and a tangent drawn to the curve from this point (see Figure 3.9).

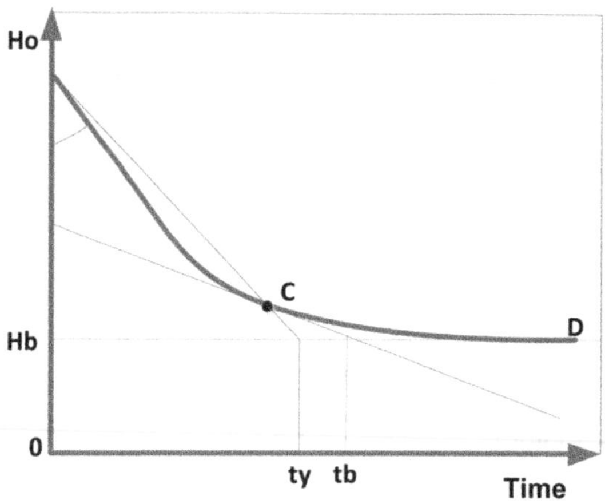

Figure 3.9. *Oltmann method*

We draw the horizontal line at point Hb corresponding to the desired sludge concentration Xb, knowing that, according to Kynch, we always have:

$$HoXo = HbXb$$

The term tb is determined graphically from the intersection of the horizontal line with ordinate Hb and the tangent at the compression point C. The area of the settling tank is given by the relation:

$$S = \frac{tb}{HoXo}$$

where tb is the time required to obtain the desired concentration.

We use the abscissa ty of the intersection point between the line with ordinate Hb and the line joining the point with ordinate Ho of the curve and the point C, called the Oltmann line (see Figure 3.9). We write (Suttill 1991):

$$S = \frac{ty}{HoXo}$$

Surface area is always expressed in $m^2.kg^{-1}.d^{-1}$ or as S (m^2)/Qe ($m^3.d^{-1}$) Xa($kg.m^{-3}$).

3.4.3. *Compaction zone*

In this zone, settling velocities are very low and correspond to the settling of the sludge. The deposited solids enter a compression phase, in which the particles are in physical contact with each other, slowly promoting the thickening process. The particles come into contact and compress the lower layers. This phase is used to determine the sludge storage height. This compression phase is primarily used for sizing the thickeners (see Volume 4, Chapter 3).

3.5. Sizing of clarifiers

There are various fundamental principles of activated sludge clarification. The hydraulics of a clarifier are a fundamental element encompassing several parameters:

– the settling area (m^2), which defines a “speed” based on the crossflow rate $m^3.m^{-2}.h^{-1}$ or $m.h^{-1}$;

– the settling height (m), which defines a clarifier volume and therefore a hydraulic retention time, as well as a temporary storage capacity for the settled sludge;

– sludge recirculation, which is carried out by pumping the scraped or suctioned sludge, aims to return it to the bioreactor where the biological treatment of the pollution takes place. This recirculation is the element through which the activated sludge tank and the clarifier are linked (see Figure 3.9). This link is expressed by the recirculation rate (recirculated flow rate expressed as a percentage of the clarified flow rate at the plant) and by the following equation:

$$Xr = ((r+1)/r)\ Xa$$

where r is the recirculation in %, Xa is the concentration of activated sludge in the biological tank and Xr is the concentration of recirculated sludge pumped from the bottom of the clarifier.

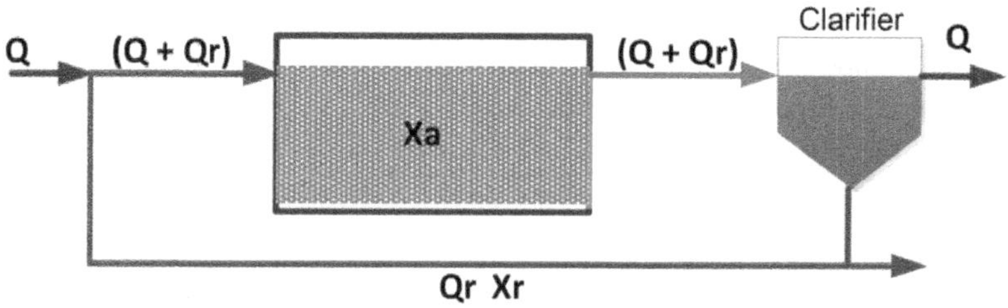

Figure 3.10. *Schematic of sludge recirculation*

Recirculation thus induces an equilibrium between the concentrations of sludge entering and leaving the clarifier. It should be noted that the "speed" of a clarifier is calculated by considering only the flow rate of the effluent to be treated, without the recirculation flow rate. The load on the bottom of the tank ($kgTSS.m^{-2}.h^{-1}$) results from the flow rate and concentration of the mixed solution. The removal of flocs present in the liquid medium is achieved using only the forces of gravity. The sedimented flakes correspond to various physical states. Examining a settling curve reveals varying particle evolution within the clarifier. The way particles evolve is specific to the physicochemical nature of the flakes. The settling velocity depends on the direction of liquid flow (horizontally, vertically, circularly, or spirally), the type of flow (laminar or turbulent), the viscosity of the medium and the shape, diameter and density of the flocs formed during flocculation. A spherical particle with density ρs settles with increasing velocity until the attractive force and the resistance force due to friction are balanced.

The settling velocity vc becomes constant and is equal to:

$$vs = \frac{g\,(\rho s - \rho)}{18\,\mu})\ d^2$$

where vs is the settling velocity of a particle ($m.s^{-1}$); g is the acceleration due to gravity (9.81 $m.s^{-2}$); ρs and ρ are the densities of the particles (flocs) and water, respectively ($g.cm^{-3}$); μ is the dynamic viscosity (1.1 x 10^{-3} Pa.s); and d is the diameter of the particles (m).

vs therefore depends on the diameter of the particles (flocs), their density and the viscosity of the water (which itself depends on the temperature).

Determining the size of the settling tanks is based on the following data:

– The volume of sludge to be treated per unit area and time ($m^3.m^{-2}.h^{-1}$), characterized by the surface hydraulic head to which the retention time is linked; this is the case for dilute suspensions.

– The mass flow of particles to settle per unit area and time ($kg.\ m^{-2}.h^{-1}$), characterizing the quantity of suspended matter to settle; this is the case for concentrated suspensions.

In both cases, test tube tests are done in order to evaluate the settling velocity.

Several empirical procedures exist for determining the surface area of a settling tank from the settling curve obtained by settling a suspension in a graduated cylinder with a capacity of at least one liter. It is necessary to work with representative samples and at a constant temperature.

The second fundamental point of clarification concerns the quality of the sludge, which must be separated from the clarified water:

– The concentration of activated sludge determines the concentration of particles upon their arrival in the clarifier. According to settling theory, the higher this concentration, the more difficult the separation of water and particles will be, as the particles will interfere with each other during their descent to the bottom. Conversely, the lower the concentration, the easier this separation will be. A minimum of 0.5 to 1 $g.L^{-1}$ of activated sludge remains necessary for optimal settling due to the minimum quantity of bacterial particles required to ensure optimal structuring (also called “flocculation”).

– The sludge index (SI) represents the capacity of activated sludge to settle rapidly. This is measured by letting one liter of fresh activated sludge (taken a few minutes beforehand from the aeration tank) settle for 30 minutes and is expressed in ml (of settled activated sludge)/l. The test reference is called “V30”. With the concentration of the sludge tested in $g.L^{-1}$, the sludge index will be given in $ml.g^{-1}$ ($ml.L^{-1}$ divided by $g.L^{-1}$). Ideally, to ensure the reliability of this index measurement, the activated sludge will have been diluted (prior to the 30-minute settling test and using clarified water from the same site) so that approximately 1 g of sludge is present in the test liter. This avoids the scaling effects associated with using only a one-liter test tube. The exact concentration will then allow correction of the value and the expression of the index in ml.g^{-1}.

The corrected volume (Vc) is equal to the sludge index (ml.g^{-1}) multiplied by the concentration of activated sludge (g.L^{-1}). This corrected volume is directly related to the maximum clarification velocity, leading to the graph presented in Figure 3.11, which involves the limiting surface hydraulic head or upward velocity ($m^3.m^{-2}.h^{-1}$). This is obtained from the ratio between the allowable flow rate ($m^3.h^{-1}$) and the passage area (m^2).

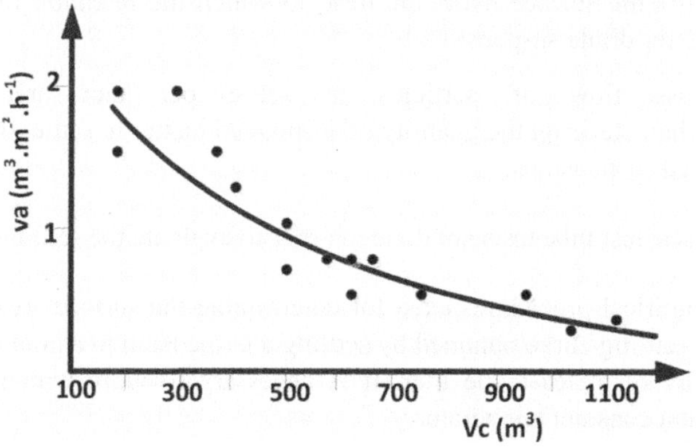

Figure 3.11. *Corrected sludge volume as a function of hydraulic load*

From the corrected volume and the mirror rate, the mass volume flow can be deduced (Qsv, in l.$h^{-1}.m^{-2}$). The sludge thickening capacity is the final parameter to consider, and it is important because the sludge recirculation must be adapted to this thickening, which varies depending on the type of sludge, its concentration at the clarifier inlet, and ultimately the concentration that must be reached at the bottom of the clarifier to ensure 100% biomass return without prolonged storage in the system (risk of denitrification for sludge with low loading rates or fermentation for sludge with medium and high loading rates).

FNDAE report 35 (Canler and Perret 2008) indicates maximum sludge retention times at the bottom of the clarifier of 45 to 60 minutes to prevent sludge re-emergence through fermentation or denitrification and recommends sludge removal by suction beyond 25 m in diameter.

The graph from the ATV-A131 F report (see Figure 3.12) illustrates the concentration limits generally considered for activated sludge (ATV-DVWK 2000).

It is important to clearly define the operating conditions of the clarifiers, in particular the following parameters:

– concentration of suspended solids entering the clarifier, average and in months (or weeks) of peak load;

– average sludge index considered, and maximum index considered over a month;

– average flow rate to be treated, peak flow rate during dry weather and maximum flow rate during wet weather;

– average and peak expected recirculation rates;

– type of sludge recovery at the bottom of the clarifier (V-shaped scraper, suction scraper, helical scraper).

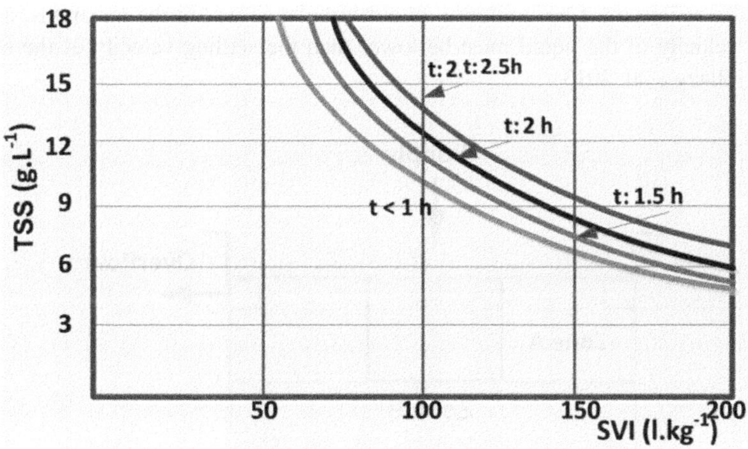

Figure 3.12. *SVI according to the concentration of the settled sludge (t (h): sludge thickening time at the bottom of the clarifier)*

3.5.1. *Different sizing approaches*

Sizing of the clarifier essentially consists of determining the maximum upward velocity of the structure. A continuously operating clarifier can be divided into four zones, as shown in Figure 3.13:

– Zone A: the effluent is clarified and passes into the overflow. This is the upper part of the clarifier from which the effluent is discharged into the receiving environment from the overflow blade, where the concentration of suspended particles is very low. In most cases, the concentrations of suspended matter in this area do not exceed 30 mg.L^{-1}. This first phase is quite short. The clarifier quickly produces a bright, clear surface effluent; that is, surface brightening.

– Zone B: the suspension has a relatively uniform solid concentration, and free or coalescent settling conditions predominate.

– Zone C: this is a transition zone where conditions evolve from free or coalescent settling to compression flow.

– Zone D: the compression of the solids forces the liquid out of the interparticle space. The surface area of the settling tank must be sufficient to allow the particle with the lowest settling velocity to reach the bottom of the apparatus. However, since the settling velocity varies depending on the zone, the volume must be large enough to allow the particle to pass from one zone to another. Similarly, the flow rate of the clarified overflow must be limited to avoid turbulence within the apparatus, and the upward velocity of the liquid must be lower than the settling velocity of the slowest particle (Blazy et al. 2013).

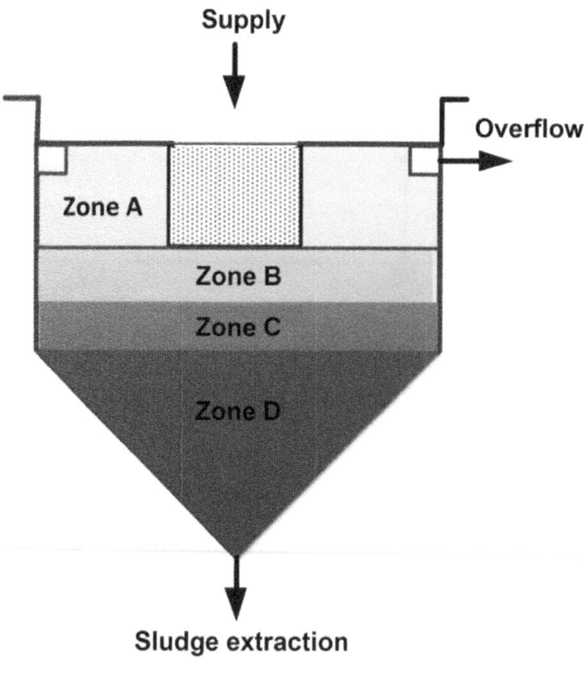

Figure 3.13. *Vertical flow settling tank*

3.5.2. *Coalescing settling (zone B, Figure 3.13)*

This phase is slightly longer. It corresponds to a constant settling of particles. The solid–liquid interface stabilizes at a uniform rate and a constant concentration.

As dilution decreases, the particles move closer together and their behavior undergoes considerable changes. The solids settle as a consolidated mass. The particles influence each other's movement. The interface between the suspension and the supernatant is now clearly visible. When the particles collide, flocculation is promoted and therefore settling. In a horizontal flow clarifier, particles can travel along a curved path, settling more quickly in the first few meters of the structure and not following the diagonal line. When they adhere to this line, the longer retention time allows for more collisions between particles and their growth. This promotes a higher terminal settling rate. In practice, clarifiers must be of a certain depth to prevent hydraulic short circuits; they are constructed at a depth of 3 to 6 m with retention times of at least two hours.

Shallow depths are exploited by incorporating a deeper zone at the clarifier inlet, which allows for the collection of heavier sludge.

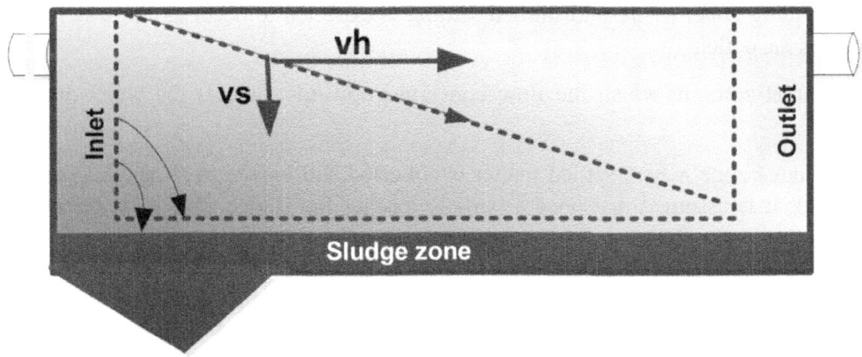

Figure 3.14. *Horizontal clarifier with inlet sludge pit*

Idealized representations of three common types of clarifiers are presented: rectangular horizontal flow, circular radial flow and upward flow.

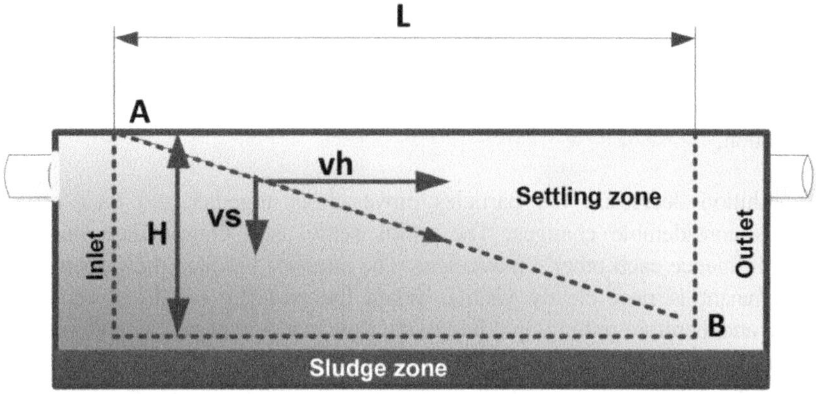

Figure 3.15. *Rectangular horizontal flow clarifier*

The ideal rectangular horizontal flow settling tank is considered to be divided into four zones (see Figure 3.15):

– inlet zone, in which momentum is dissipated, and flow is established in a uniform forward direction;

– settling zone, where undisturbed settling is expected to occur as the water flows towards the outlet;

– outlet zone, in which the flow converges upwards towards the weirs or wash plants;

– sludge zone, where settled matter is collected and moved to sludge hoppers for removal. It is assumed that once a particle reaches the sludge zone, it is effectively removed from the flow.

The critical particle in the settling zone of an ideal rectangular settling tank is the one that enters from the top of the settling zone (point A) and settles at an optimal velocity to reach the sludge zone at the outlet (point B). The two velocity components of such a particle are vh for the horizontal flow and vs for the settling velocity.

The time t required for the particle to settle is given by:

$$t = H/vs = L/vh \text{ and } vs = H/t$$

where vh is the horizontal velocity.

Thus, $t = V/Q$, where V is the volume of the clarifier.

$vs = H\ Q/V$, $vs = Q/S$ and $vs = Q/L\ l$

where Q is the flow rate, and L, l and H are respectively the length, width and depth of the clarifier. Since the surface area of the clarifier S is L l, then:

$vs = Q/S$

According to this relationship, the slowest-settling particles, which we might expect to be completely removed in an ideal clarifier, would have a settling rate of Q/S. This parameter, called the surface loading or surface velocity, is a fundamental parameter governing the performance of the settling tank.

This relationship also implies that the settling efficiency is independent of the tank depth – a condition that is only true if the scraping speed is low enough to ensure that the deposited materials are not scoured back into suspension at the bottom of the structure.

A similar analysis of an ideal circular radial flow settling tank is summarized in Figure 3.16.

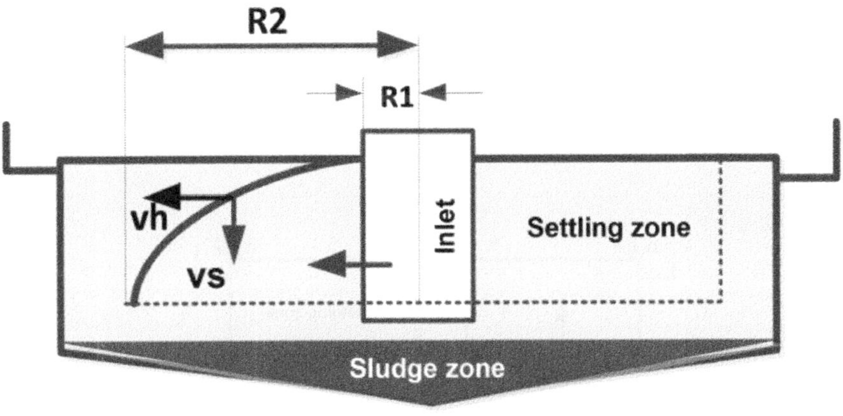

Figure 3.16. *Circular clarifier*

The retention time d is given by:

$t = H/vs$ and $t = \int_{R1}^{R2} dr/vs$

Therefore:

$$vs = Q/2\pi H$$

which then becomes:

$$t = \frac{2\pi H}{Q}\int_{R1}^{R2} r\,dr = \frac{\pi(R2^2 - R1^2)H}{Q}$$

We also have:

$$\pi (R2^2 - R1^2) = S$$

and:

$$\frac{SH}{Q} = \frac{H}{Qvs}$$

Finally, we find that vs = Q/S.

In an ideal upflow settling tank (see Figure 3.17), it is evident that a particle will only be removed if its settling velocity exceeds the upflow velocity (vup) of the water. In this case, the minimum upward flow velocity is given by the flow rate divided by the tank surface area (Q/S), so, once again, the minimum settling velocity for a particle to be removed is vs = Q/S.

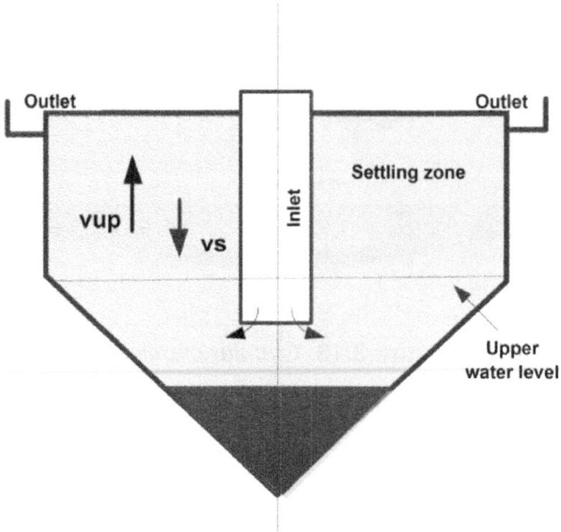

Figure 3.17. *Upflow conical clarifier*

3.5.3. *Hindered settling zone and sludge blanket clarifiers (zone C, Figure 3.13)*

The hindered settling zone is the layer that separates the two zones of clarification and compression. As the particle concentration in a suspension increases, a limit is reached when the particles are so close to each other that they no longer sediment independently, and the fluid velocity fields displaced by the particles overlap. The particles settle in groups, meaning they interact with each other. In a cylinder sedimentation test at a sufficiently high concentration, a clear liquid supernatant and a turbid suspension clearly separate.

There is also a net upward flow of liquid displaced by the settling particles. This results in a reduced particle settling velocity, and this effect is known as hindered settling. This situation occurs when the particle concentration is so high that the entire suspension tends to settle as a "blanket." This is referred to as zone settling, because it is possible to distinguish several distinct zones, separated by discontinuities in concentration. Within this zone, the flocs generally have the same solid structure, and all settle at roughly the same rate.

Zone settling is important in the final design of clarification tanks for activated sludge processes.

Upflow clarifiers often operate with a high concentration of suspended sludge. A suspension will only be retained in this type of clarifier if the settling velocity at the interface of the suspended sludge, vi, is equal to the upflow velocity, vf. These clarifiers maintain a high concentration of suspended solids and promote collisions between particles and their agglomeration. These sludge blanket clarifiers help retain a large number of very fine particles that would otherwise have been carried over into the effluent.

Zone settling is described according to the Vesilind relation (Vesilind 1968):

$$vszone = vsmax\ e^{-rh\ X}$$

with:

– vszone: zone settling rate ($m.h^{-1}$);

– X: TSS concentration ($g.L^{-1}$);

– vsmax: maximum settling velocity ($m.h^{-1}$);

– rh: zone setting parameter ($L.g^{-1}$).

During this phase, a balance between gravitational forces and frictional forces means that the setting rate is the same for all particles in the region (Valle Medina 2019).

3.5.4. *Sludge compaction zone (zone D, Figure 3.13)*

Settling continues until the sludge blanket is reached, where it compacts under its own weight and the action of the scraper. In this compaction zone, the sludge is treated as a network of interlocking particles, not as a fluid. In this zone, the particle density is highest, and the particles accumulate in successive layers. Each identified zone represents a different settling regime and is rather close to a thickening stage.

This regime was initially defined as the concentration range where the weight of the particles is no longer entirely supported by hydrodynamic forces. Each layer of solids would then provide some mechanical support to the layers above. Beyond a critical suspended solids concentration (SS crit equivalent to X crit), the sludge enters a compression regime: settling then corresponds to the compaction of flocs under the effect of their own weight, with the appearance of solid stress.

The compression zone forms the lower part of the secondary clarifier where the sludge is stored. Very high sludge concentrations develop as the settled particles approach the tank bottom. The sludge mass stabilizes within the matrix. This is known as compression settling or compaction. This phenomenon occurs when deposited solids are compressed under the weight of overlying solids, progressively decreasing porosity and reducing the water content of the matrix. Compression settling is important in gravity thickening processes. It is also important in activated sludge clarifiers, where the sludge must be thickened for recycling to aeration tanks and for the removal of a fraction of the sludge. In this zone, the floc structure is properly consolidated and has a higher density, and consequently, settling is greater.

This phenomenon is described by the compression function dcomp, according to the equation:

$$\mathrm{dcomp} = \frac{\rho b\,\lambda}{g(\rho b - \rho e)}\,\mathrm{vhs}(X)$$

with:

– dcomp: compression function ($m^2.s^{-2}$);

– g: acceleration due to gravity (9.81 $m.s^{-2}$);

– λ: solid stress parameter according to Torfs et al. (2017) ($m^2.s^{-2}$);

– ρb: sludge density ($kg.m^{-3}$);

– ρe: water density ($kg.m^{-3}$).

After a certain period of settling, zones B and C gradually disappear, giving way to the compaction zone. The evolution of the height of the A–B and then A–D interfaces over time is clearly seen in the test tube tests.

Compaction is accompanied by the ascent of the liquid by percolation through the sludge matrix. Compaction can be measured by the difference between the solid fraction obtained at the bottom of the clarifier and the initial solid fraction.

The following equations define the solid fraction, compaction and compaction rate, respectively (Imam et al. 1983):

$$X = \text{mass of dry sludge/total mass of the sample} \cdot 100$$

where X is the solid mass fraction (% $kg.kg^{-1}$).

Variation in the compaction of a settling tank = (% $kg.kg^{-1}$) final – (% $kg.kg^{-1}$) initial

Settling tank compaction rate = compaction/Δt

Compaction is expressed in (% $kg.kg^{-1}$) and Δt is the time required to achieve compaction, expressed in hours.

Settling theory predicts that, in the case of ideal settling, the main design parameter to consider is surface loading (Q/S) because it represents the critical particle settling rate. Coalescent settling requires that an adequate depth, H, or retention time, t, be provided in order to allow flocculation.

In practice, it is not always possible to assume a uniform flow distribution due to density currents and insufficient dissipation of the impulse at the inlet. Because of all these effects, the surface charges and retention times derived from the theory must be multiplied by a safety factor often between 1.7 and 2.5. In the case of hindered settling (zone settling), it has also been shown that surface loading is the main parameter to consider during design. The majority of the theoretical development therefore applies to the design of clarifiers in these two zones.

Designing clarifiers for a given flow rate Q involves choosing a surface head Q/A, from which the surface area of the structure can be calculated. The relationships between the various parameters involved can be expressed as indicated below.

For Q in $m^3.h^{-1}$ and S in m^2, the upward velocity ($m.h^{-1}$) is given by:

$$vs = Q/S$$

Sludge indices (SI), suspended solids concentrations in sludge from aeration tanks (Xa), are used to calculate the maximum permissible surface hydraulic rate (Ch) acceptable for a good settling.

The methods recommended by the French National Research Institute for Agriculture, Food and Environment (INRAE) (ATV-DVWK 2000; Canler and Perret 2005) are the most commonly used.

Knowing the sludge concentration (X) ($g.L^{-1}$) and the sludge index (SI) ($ml.g^{-1}$), Ch can be deduced from the relationship:

$$\text{INRAE: Ch } (m.h^{-1}) = 2.65 \text{ Exp } (-0.00193 \text{ Xa SI})$$

$$\text{ATV: Ch } (m.h^{-1}) = 500/\text{Xa SI}$$

The retention time is: t (h) = SH/Q, where H is the total height of the unit, and S is the surface of the clarifier.

Numerous equations have been proposed, such as that of Vesilind (1968) (see Figure 3.17):

$$\text{Ch } (m.h^{-1}) = 9.6753 \ e^{-0.434 X}$$

Cho et al. (1996) propose a similar one:

$$\text{Ch } (m.h^{-1}) = 25.905 \ [(1-0.049X)^4]/X$$

A concentration of sludge of 4 $g.L^{-1}$ would settle at a rate of 2.70 $m.h^{-1}$, whereas if the concentration were to decrease to 3 $g.L^{-1}$, the rate would increase to 3.42 $m.h^{-1}$. Since the upward velocity would be slightly lower than the settling velocity, the surface velocities would be 3.1–3.2 $m.h^{-1}$ for the second case. Tests have been carried out in batch mode and do not take into account turbulence and dynamic movements within the clarifier. Cho et al. (1996) propose numerous relationships depending on the operating conditions (see Table 3.3).

Relationships vs TSS concentration
$Ch = 257.8\ X^{(-3.245)}$
$Ch = 17.89\ EXP(-0.468X)$
$Ch = 41.88\ EXP(-0.327X)/X$
$Ch = 23.50(1 - 0.057X)^3/X$
$Ch = 8.448(1 - 0.055X)^{4.65}$

Table 3.3. *Settling rate relations (according to Cho et al. (1996))*

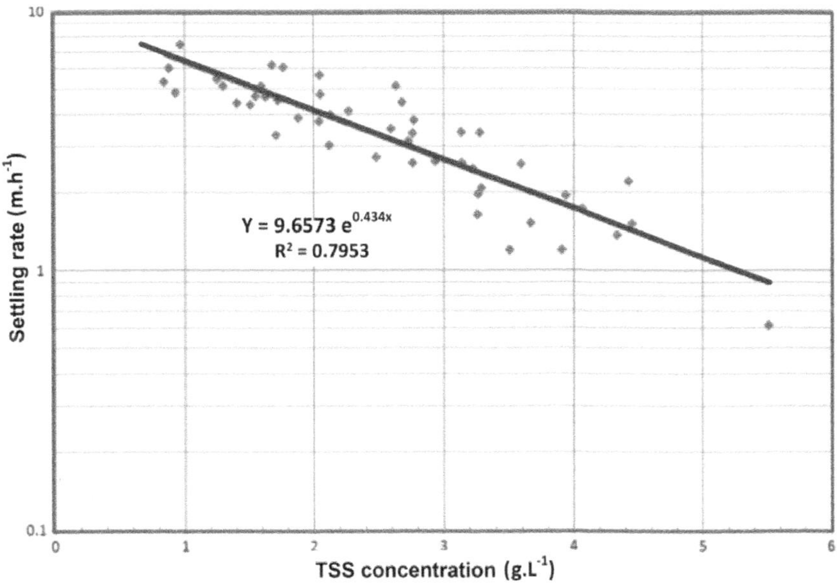

Figure 3.18. *Settling rate as a function of TSS concentration*

There are two types of equations in the literature: Ch = a exp(-nX) and Ch = a' Xi^{-n}, in which a, a' and n are activated sludge constants.

3.5.5. *Relationship between settling rate and height of equipment*

The height of the structure corresponds to the sum of the heights of the four layers within a clarifier.

The basis for determining a clarifier size is as follows (see Figure 3.19):

– Zone 1 (H1): clear water, generally at a height between 0.5 and 1 meter.

– Zone 2 (H2): separation zone. Its height depends directly on the flow velocity and the sludge index.

– Zone 3 (H3): storage zone, which allows for sludge storage during peak periods (dry weather or rain), and which involves the recirculation rate.

– Zone 4 (H4): thickening zone, related to the sludge thickening time within the clarifier itself.

The total height H, taken into account when determining the size of the structures, is measured at two thirds of the radius of the foundation slab.

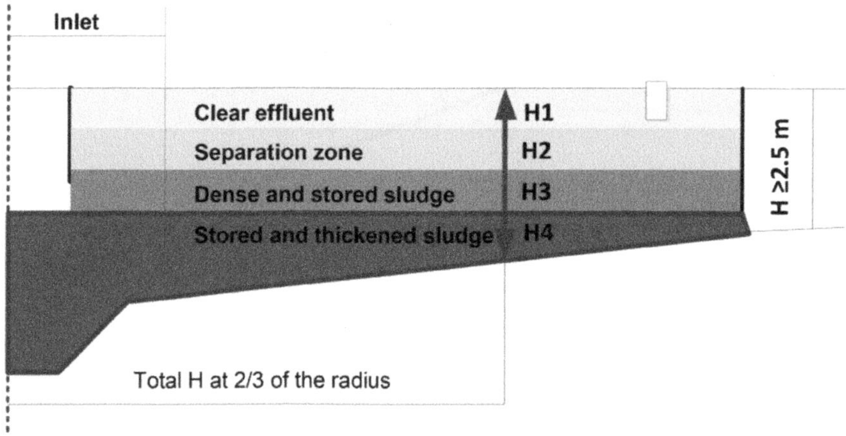

Figure 3.19. *Diagram describing the different zones of a clarifier*

In all cases, for a given corrected volume, a greater water height in the clarifier can accommodate higher peak clarification rates thanks to:

– a greater storage height and temporary thickening of the sludge, thereby allowing greater particle separation and limiting particle movement in the event of a rise in the sludge layer (as in the case of peak rainfall);

– reduced sensitivity of the sludge blanket level (the layer of settled sludge at the bottom of a clarifier) to inlet turbulence in the clarifier thanks to more space between the bottom of the diffuser and the sludge blanket level.

It is very important to clearly differentiate between the maximum velocities that can be accepted for a limited peak flow of 4 hours and the maximum velocities that must be sustained without sludge release for 24 hours; these latter velocities are, of course, lower and will be used to guarantee the performance of the structure.

Based on the technology descriptions, the criteria for choosing one technology over another are:

– Low upward velocity and low bottom loads: a shallow clarifier is suitable.

– High bottom load: a clarifier with better sludge removal is recommended.

– High foundation velocity and load: a clarifier with a large height is required.

	Scraping mechanism			
	Scraped	**Suction**	**Helical**	**Suction (perforated tube)**
Average foundation load ($kgTSS.m^{-2}.h^{-1}$)	1.5–3	3–6	5–7	7–9
Average recirculation rate (% clarified flow rate)	100–150		105	100–120
Maximum foundation load ($kgTSS.m^{-2}.h^{-1}$)	5	7.5	9	12
Minimum recirculation rate (% clarified flow rate)	100		80	80

Table 3.4. *Recirculation rates and base loads*

Manufacturers most often offer, in the case of activated sludge with prolonged aeration, an upward velocity of 0.4–0.6 $m.h^{-1}$ for determining the size of the structure. This upward velocity depends on the sludge characteristics (concentration and settling capacity).

The clarifier is supplied with sludge from the aeration tank; its characteristics have a significant impact on the sizing of the structure and its performance.

The main parameters to consider are the following:

– the concentration of sludge from the aeration tank obtained by measuring suspended solids (TSS);

– the sludge's ability to settle (seedling test – sludge index).

The sludge index is the ratio of the volume of sludge settled after 30 minutes to the number of TSS in the test tube. It expresses the volume occupied per gram of TSS ($ml.g^{-1}$ of TSS). Values above 200 ml indicate the presence of filamentous bacteria in the sludge.

Like any test, the sludge index protocol must be strictly followed. Its implementation most often requires pre-diluting the sludge sample with wastewater from the treatment plant. Its validity is only guaranteed if the settled volume after 30 minutes is less than 300 ml. By maintaining a settled volume of less than 300 ml after 30 minutes, the sludge index can be used to measure the settleability of the sludge.

The Mohlmann index (undiluted test) is normally used to address sludge thickening problems but cannot be used for sizing clarifiers (determining the limiting upward velocity).

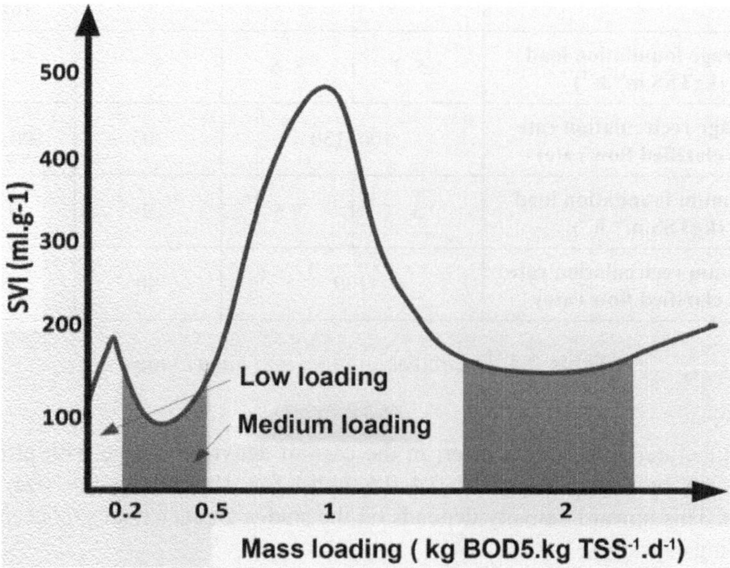

Figure 3.20. *SVI as a function of mass loading*

Other parameters associated with the sludge can influence the operation of the clarifier, such as:

– Dissolved O_2 concentration: this can be disadvantageous if there has been over-aeration with pure oxygen; a strong insufflation of air into the biological tank leads to degassing at the inlet.

– Air bubbles at the outlet of the aeration tank due to insufficient degassing.

– Nitrate concentration (N-NO_3^-): uncontrolled denitrification can begin and lead to the degradation of the discharge due to sludge seepage.

– The retention time of the sludge at the bottom of the structure.

At the same upward velocity, cylindrical structures allow for better retention of suspended solid flows, mainly due to the following factors:

– the lower flow rate of water over the nappe;

– a more homogeneous distribution of the sludge.

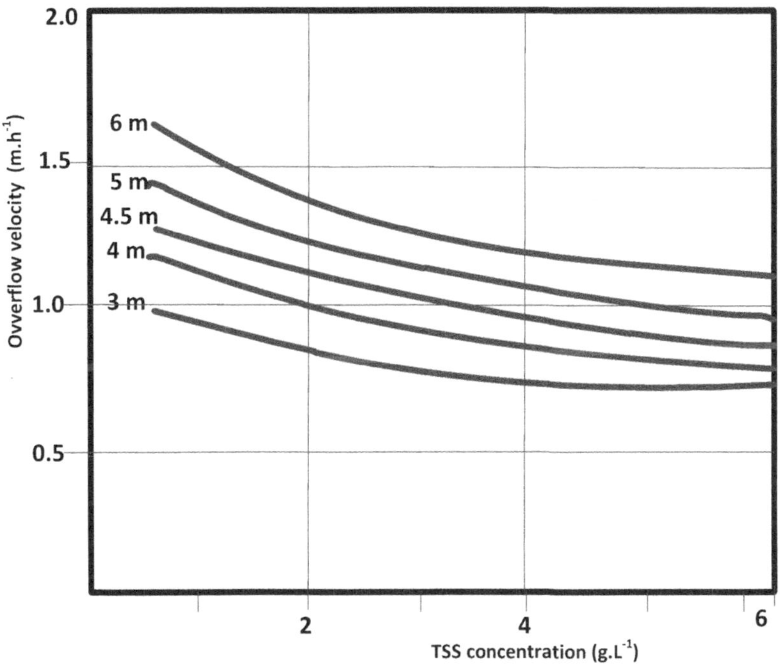

Figure 3.21. *Influence of clarifier height on overflow velocity and sludge concentration (WEF 2005)*

3.6. Characteristics of classic clarifiers

The clarifier must fulfill several essential functions:

– the separation and retention of the maximum number of suspended particles (TSS);

– thickening (sludge concentration);

– temporary storage of sludge from the aeration tank during periods of increased, localized hydraulic loads.

There are two main, distinct clarifier configurations, differing in their supply type and operating method:

– Rectangular structures with a sludge introduction device at one end, with a length-to-depth ratio greater than 6. The structure's width is constant along its entire length, promoting a stable horizontal component for the flow.

– Cylindrical or cylindro-conical structures with central feeding via the Clifford system. This configuration shortens the path of the sludge to be recirculated (limiting retention time) and can enhance the filtering action of the sludge blanket. The horizontal velocity of the water decreases during transit. The structures have a radius-to-height ratio of less than 5.

3.6.1. *Horizontal clarifiers*

In an ideal settling tank with horizontal or radial flow, particles with settling velocities < Vs can still be partially removed, but not in an ideal upwelling tank. Settling is based on Hazen's loading theory, according to which the smallest particle to be removed is the one that can be deposited along its path. In a horizontal clarifier, for a length L and a water depth H, the smallest particle that will have time to settle is determined by the settling velocity (vs):

$$vs = v\frac{H}{L}$$

where v is the (horizontal) flow velocity in the clarifier, with v= vh.

Thus, the sizing of a horizontal clarifier (also called a channel clarifier) must ensure that the flocs entering the clarifier settle at a sufficient speed to reach the bottom. The schematic longitudinal cross-section of a tank (see Figure 3.22) highlights the trajectories of the smallest particle and of any particle, along with their respective settling velocities. Thus, for a floc to settle, it must reach the clarifier's bottom before being discharged, which is expressed by the following relationship:

$$t = \frac{L}{v} \geq tc = \frac{H}{vs}$$

where t is the passage time (h) in the clarifier, vs is the settling velocity of the flocs or particles ($m.h^{-1}$), H is the total height (m), L is the length L (m) of the clarifier (up to the outlet), v is the flow velocity ($m.h^{-1}$) and tc is the settling time of the flocs (h).

For example:

$$vs \geq \frac{H.v}{L} = \frac{Q.H}{L.H.l} = \frac{Q}{S}$$

In theory, the particles will only be retained in the clarifier if vp $\geq$ v.

$$vs = \frac{v.H.l}{l.L} = v\frac{H}{L}$$

In Figure 3.22, Q is the processing capacity ($m^3.h^{-1}$); v is the input velocity ($m.h^{-1}$); H, l and L are the height (m), the width (m) and length (m) of the clarifier, respectively; vs is the settling velocity of the flocs ($m.h^{-1}$); and S is the fluid surface of the clarifier (m^2).

The equation above means that the processing capacity is proportional to the fluid surface area of the clarifier. The smallest flocs should settle as long as their settling velocity is greater than or equal to Q/S. It is therefore important to control this parameter by obtaining the best possible flocculation, which leads to flocs with the largest possible diameter and results in an increase in Stokes velocity.

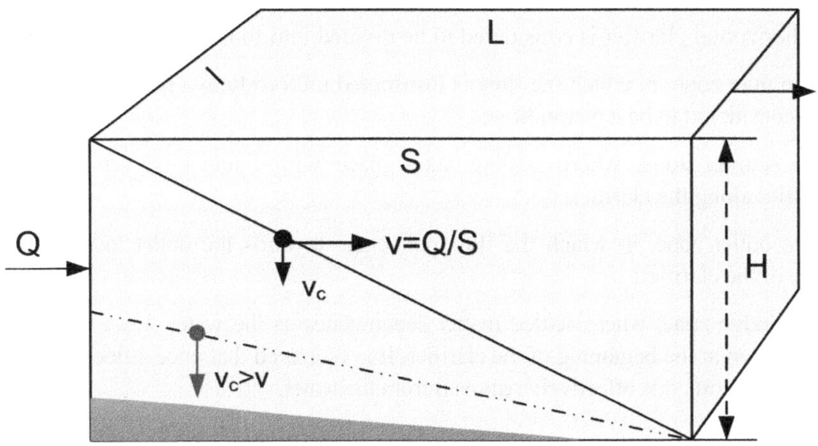

Figure 3.22. *Floc settling velocity and flow velocity in a horizontal clarifier*

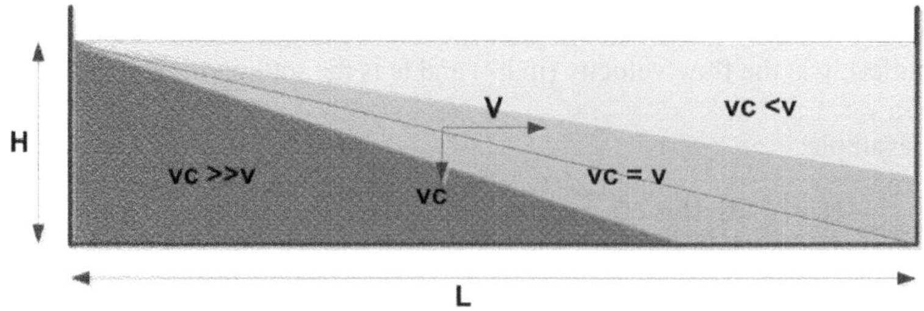

Figure 3.23. *Sedimentation profile as a function of the settling velocity and the flow velocity*

If the floc settling velocity is lower than the flow velocity (surface velocity), the flocs settle at the bottom of the clarifier. Turbidity may be reduced (see Figure 3.23). When the floc settling velocity is higher than the flow velocity, the flocs settle very quickly at the top of the clarifier – this is the case for water with a high suspended solids content.

When the floc settling velocity is close to or equal to the flow velocity, the flocs settle as the water flows and are dispersed over two-thirds of the clarifier's length. This is the case for typical water with a turbidity ranging from 20 to 100 NTU. The turbidity in settled water is less than 1 NTU.

A horizontal clarifier is considered to be divided into four zones:

– an inlet zone, in which the flow is distributed uniformly in a horizontal direction and is considered to be a piston flow;

– a settling zone, where settling takes place with liquid/solid separation that intensifies along the clarifier;

– an outlet zone, in which the flow converges towards the outlet located on the surface of the clarifier;

– a sludge zone, where settled matter accumulates as the water flows with a high concentration at the beginning of the clarifier. It is estimated that once a floc reaches the clarifier's bottom, it is effectively removed from the water.

Therefore, the length of the clarifier takes this into account so that even the smallest settleable particles can be removed from the water. The length can extend up to 80 m, while the maximum observed width is 20 m.

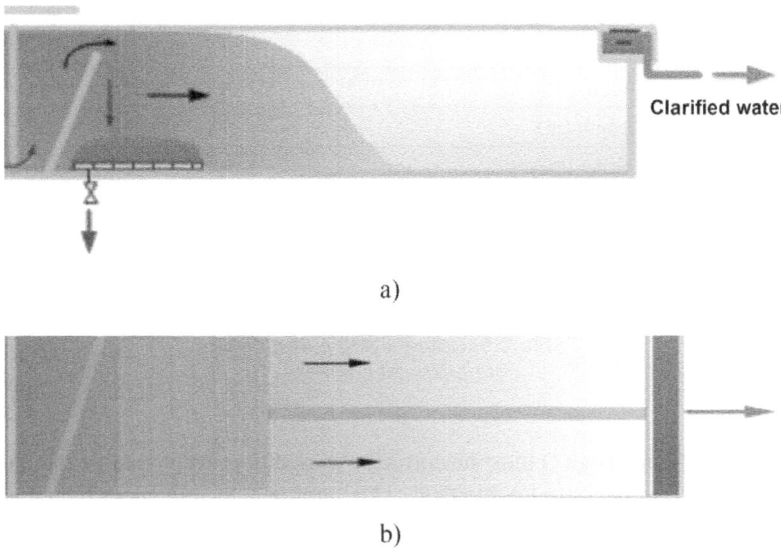

Figure 3.24. *Representation of the operation of a horizontal clarifier with: a) side view; b) top view*

3.6.1.1. *Design*

One of the main parameters to consider is the flow velocity ($m.h^{-1}$) because it is linked to the floc settling velocity. Indeed, the settling efficiency is theoretically independent of the tank depth, provided that the velocity at the surface (flow velocity) is sufficiently low so that the settled flocs are not resuspended by turbulence. This is why precautions are taken by operating with variable depths between 3.5 and 5 m.

The flow velocity ($m.h^{-1}$) is given by: $v = \frac{Q}{S}$.

The retention time (h) is given by: $t = \frac{S.H}{Q}$.

Q is the inlet flow rate ($m^3.h^{-1}$), S is the fluid surface area (m^2) and H is the water level (m).

Retention times in horizontal clarifiers vary between 2 and 4 hours, although some clarifiers have retention times of up to 12 hours.

Parameter	Size
Flow velocity ($m.h^{-1}$)	0.5–1.2
Width (m)	5–20
Length (m)	20–80
Water height (m)	3–5
Retention time (h)	2–4
L/H ratio	< 18
L/l ratio	3–5

Table 3.5. *Some sizing parameters of horizontal clarifiers*

The inlet velocity ($m.h^{-1}$) must be correctly chosen to prevent resuspension of the settled flocs due to oversized equipment. This requirement influences the choice of length/width and height/width ratios for this type of clarifier. It is based on the relationship:

$$\text{vmax} = \frac{Q}{H.l} = \frac{L}{t}$$

where l is the width of the clarifier (m). In practice, the L/l ratio varies between 3 and 5, with a value of 4 being recommended. The performance of horizontal clarifiers is sensitive to temperature differences due to the low flow velocity (< 0.5 $m.h^{-1}$) which characterizes these structures, particularly in hot weather. A temperature difference of only 1°C is sometimes enough to cause a decrease in efficiency due to changes in density.

3.6.1.2. *Implementation*

It is clear that ideal settling is not possible to achieve in a factory due to variations in flow rates, variations in sludge loads, the difficulty of obtaining a homogeneously distributed flow, short circuits caused by sludge deposits, etc.

Nevertheless, manufacturers implement all technical means to achieve the best liquid/solid separation. The clarifier is sized so that even the smallest flocs can be retained, based on contact times that ensure very high floc removal. A well-sized horizontal clarifier has an outlet concentration of suspended solids below 30 $mg.L^{-1}$.

The optimal flow velocity is less than 0.5 $m.h^{-1}$. When this velocity rises to 1 $m.h^{-1}$, yield decreases and around 98.5–99% settles for a load of 500 $mg.L^{-1}$.

Clarifiers must include submerged inlet ports in their feed system to ensure a stable and uniform flow rate. The clarified water is collected in perforated pipes located at the end of the clarifiers or in troughs covering approximately 10–20% of the surface area.

There are two types of horizontal clarifiers with scraper blades: traveling bridge clarifiers and chain clarifiers. The traveling bridge moves back and forth and performs the scraping action in a counter-current motion. During the scraping phase, the speed cannot exceed 3 cm/s, which limits the scraping devices to 60 m.

Chain clarifiers allow for the continuous scraping of sludge and floating matter by a series of scrapers mounted between two parallel chains that rotate along the vertical walls of the settling tank.

Horizontal clarifiers are advantageous when ground space is available. However, this is becoming increasingly rare, meaning that it is necessary to consider more compact clarifiers while maintaining equivalent performance.

Horizontal clarifiers can exhibit reduced performance if the laminar flow is not evenly distributed, creating hydraulic short circuits that disrupt their proper operation. Furthermore, sludge extraction must be carefully planned across the entire surface of the tank bottom to prevent accumulations in various areas of the clarifier, which would lead to floc leakage. This would reduce the turbidity at the outlet. Finally, the concentration of the extracted sludge is relatively low, thus requiring a larger thickening structure.

For all these reasons, the production of circular clarifiers and lamellar clarifiers has developed, with the introduction of new compact and efficient technologies to the market.

3.6.2. *Cylindrical clarifiers*

Let ro be the radius of the central inlet, R that of the clarifier with height H, and vo and vR the horizontal velocities of the water at the inlet and periphery of the settling tank, respectively. If we consider a settling particle, its horizontal velocity (vh) being that of the water at the point considered, at a distance r from the axis of the tank, we can write:

$$vh = dr/dt = vo\ ro/r$$

Its vertical velocity (vv) is constant; that is:

$$vv = dz/dt$$

Integrating these equations leads to:

$$2 \text{ vo ro t} = (r^2 - ro^2)$$

$$z - zo = vh\ t$$

By writing that the flow introduced into the device is:

$$Q = 2\ \pi \text{ ro h vo}$$

this becomes, replacing vo with its value:

$$Q = 2\pi \text{ ro H } (\frac{r^2 - ro^2}{2 \text{ ro t}})$$

And by writing:

$$t = \pi \text{ H } (\frac{r^2 - ro^2}{Q})$$

it becomes:

$$z - zo = \pi \text{ H vv}(\frac{r^2 - ro^2}{Q})$$

The most unfavorable capture case is when particles are introduced at the surface, for which we must have Oz = 0 and z = H for r = R.

Thus:

$$v = vv = \frac{1}{\pi}(\frac{Q}{R^2 - Ro^2})$$

or even:

$$v = \frac{Q}{S} \quad \text{v must be} \leq \text{vs (settling velocity)}$$

The most common clarifiers are composed of the following parts:

– The tank itself, the dimensions of which (surface area and depth) must allow for the retention of suspended solids (settling and thickening zone) and the temporary storage of sludge.

– The supply inlet (Clifford system) allows for significant dissipation of hydraulic energy and promotes even distribution throughout the structure. Its dimensions take into account all flow rates passing through this structure (water to be treated and recirculated sludge). A velocity lower than 2.5 cm. s^{-1} (< 90 $m^3.m^{-1}.h^{-1}$) has to be maintained. The area considered is that located between the central sludge supply shaft and the Clifford skirt. The immersion depth of the skirt must be between the midpoint and the lower third of the structure's height to avoid turbulence at the surface. For shallow structures, a minimum height of 1 m between the bottom of the Clifford skirt and the sludge recirculated at the bottom of the structure must be maintained. The presence of a deflector prevents the sludge exiting at the outlet from being short-circuited.

– The clarified water collection chute.

– The peripheral traction scraper for circular settling tanks, which consists of a mobile bridge functioning as a walkway, a unit with a rotating platform, a central deflector, a traction carriage, a control unit, a dirty scum container, a skimmer blade, a series of sludge scraper blades, a perimeter deflector and a weir. The rotating platform unit is mounted in the center of the tank and is equipped with axial bearings and a double-brush collector-distributor ring. The traction trolley and control unit are positioned at the walkway periphery. The wastewater inlet is central, but its flow is diverted by the deflector to optimize settling.

The settled sludge is collected and conveyed to the center of the tank by the clarifiers.

The skimmer blade and the self-discharge function are designed to remove all floating materials and substances such as oils and surfactants.

The equipment used for sludge removal has several functions:

– Concentrate the sludge in order to limit the volumes to be pumped during recirculation and extraction; it should be noted that extractions can also be carried out directly in the aeration tank, the extracted sludge then having lower but constant concentrations, which facilitates the adjustments of the thickening and dewatering station, as well as control of the sludge age;

– Limiting the sludge retention time: this depends on the recirculated flow rate, the thickening factor t (settling rate), and also the homogeneous re-entry of the sludge blanket. This re-entry will be ensured by a simple scraping system for structures with a diameter less than 20–25 m; for structures with a larger diameter, evacuation by a mixed suction-scraping system or by scraping systems should be considered.

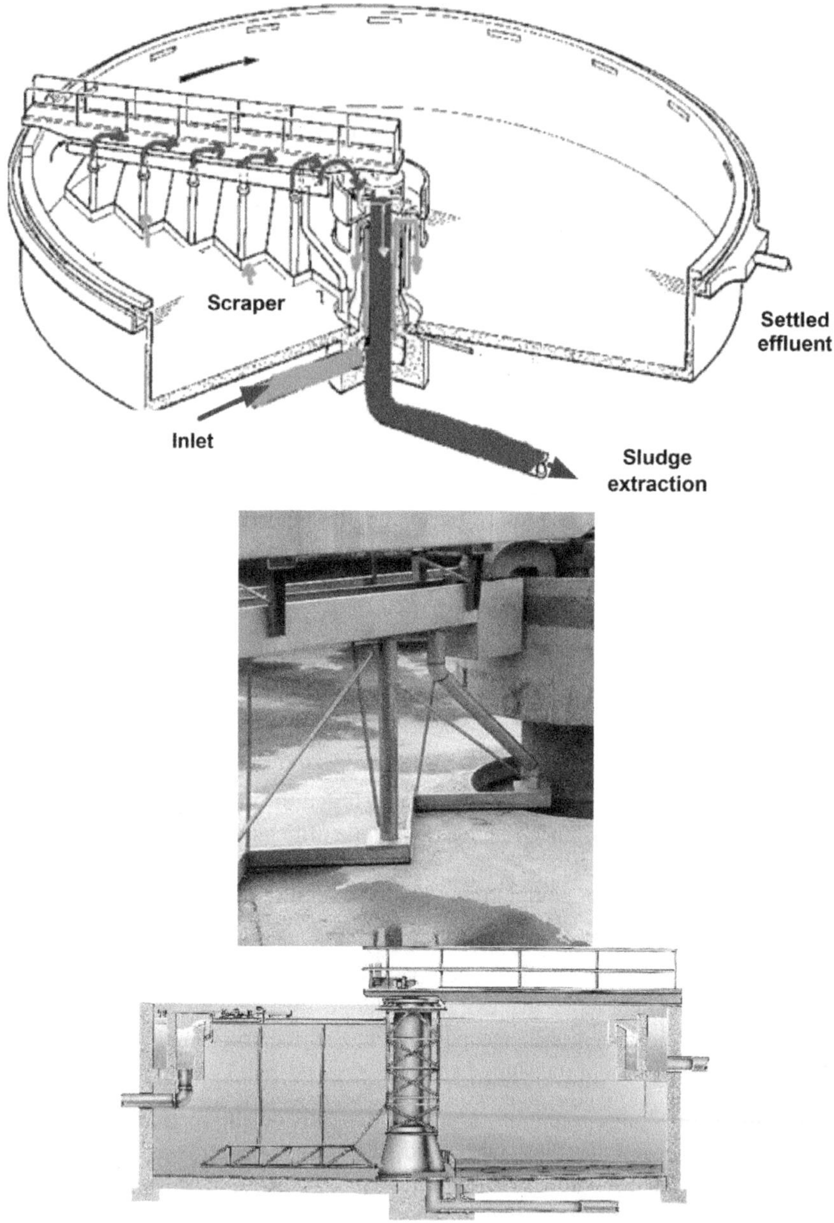

Figure 3.25. *Diagram of a cylindrical clarifier*

3.6.3. *Other technologies*

Other equipment or configurations are available. They are characterized by:

– the supply and recovery of clarified water at the periphery of the structure for the Rimflo process;

– a different design for the supply inlet, called a Coanda tulip;

– the presence of a deflector between the Clifford outlet and the sludge recovery well to reduce short circuits and therefore the dilution of the sludge to be recirculated or extracted;

– the installation of a clarified water collection chute located away from the structure edge to avoid it being situated upstream where the momentum of suspended solids is greatest;

– two clarifier stages, the second of which has the function of refining suspended solids and tertiary dephosphatation with the addition of chemical reagents (called clariflocculation): this configuration allows for a higher maximum upward velocity during the first stage. This makes it possible to design the secondary clarifiers with higher upward velocities.

3.6.4. *General design of conventional clarifiers*

Designing and determining the dimensions of a clarifier should be based on principles of settling and hydraulic design. The factors taken into account are the hydraulic head, the settling velocity and the concentration of suspended solids in the inflow.

3.6.4.1. *Surface area*

The main parameter traditionally used when determining the surface area of clarifiers is the surface hydraulic head (or upward velocity or Hazen velocity). It is expressed in $m^3.m^{-2}.h^{-1}$ or $m.h^{-1}$. It is calculated based on the peak hourly flow rate of the water supplying the structure, excluding the recirculated sludge flow rate, which is adjusted to the structure's effective surface area. This surface area (or the surface at the water surface) corresponds to the surface area of the water body without taking into account the Clifford skirt and the clarified water recovery chute when it is located outside the structure.

The limiting upward velocity depends directly on the settling characteristics of the sludge: the concentration in the aeration tank and the sludge index.

We define a corrected volume (Ve) with the relation:

Corrected volume ($ml.l^{-1}$) = sludge index ($ml.g^{-1}$) x concentration ($g.l^{-1}$)

For conventional cylindrical clarifiers with a diameter of less than 20 m and shallow depths (< 2.5–3 m), we propose:

Maximum upflow rate ($m.h^{-1}$) = 2.56 Exp (–0.00193 corrected volume)

3.6.4.2. *Depth*

The minimum depth of the structure (small structure) should be approximately 2.5 to 3.5 m to allow for thickening.

Height = suspended solids concentration in the activated sludge tank x IB/1000

The sludge accumulation depth should then be approximately 1–1.3 m, with a buffer zone of approximately 0.5 m before discharge. Greater depths allow for potential sludge storage in the clarifier during peak hydraulic load periods; the sludge flow towards the structure is momentarily higher than the sludge flow recirculated to the aeration tank. If the clarifier allows for sludge storage, the release of sludge from the aeration tank is limited to a decrease in concentration by around 1 $g.l^{-1}$ in this tank, with good quality sludge (IB = 100 $ml.g^{-1}$). The permissible upward velocity during the hydraulic peak is then calculated taking into account the lower concentration.

For communities equipped with separate sewer systems with high peak coefficients (> 2) but short retention times, the release of water from the aeration tank will be minimal and will not allow for a decrease in concentration within the aeration tank. The upward velocity based on the peak flow will be calculated using the nominal concentration of the tank. In this case, if the duration of the hydraulic peak is known, the surge will lead to an increase in the sludge blanket, which can be retained in the clarifier by implementing a greater height for the structure, hence providing higher upward velocities.

The height (in m) of the area intended for sludge storage is calculated using the following formula:

$$h = \frac{\Delta X \text{ Vba IB}}{1{,}000 \text{ S}}$$

with:

- ΔX: lowering of the concentration in the aeration tank;
- Vba: volume of the aeration tank;

– SI: sludge index;

– S: clarifier surface area.

The Vba/S ratio for prolonged aeration is of the order of 10, which gives a height of 1.5 m for the area intended for sludge storage for a sludge index of 150 $ml.g^{-1}$.

Since surface area and depth are related, greater depth allows for storage during peak hydraulic periods. Depending on the duration of the peak, the parameter influencing the surface area varies. For long peak durations, the decrease in concentration in the aeration tank resulting from this hydraulic surge leads to an increase in the upward velocity.

3.6.4.3. *Recirculation rate*

The recirculation rate affects the thickening factor, which depends on the sludge index. Thus, a high recirculation flow rate leads to a low concentration of recirculated sludge; conversely, a low recirculation rate prevents the sludge from being concentrated beyond a certain maximum concentration. This maximum concentration of recirculated sludge is given by:

Xr max = 1,000 to 1,200/SI

It would take 1,000/SI for SI < 200 $ml.g^{-1}$

and 1,200/SI for SI > 200 $ml.g^{-1}$.

The recirculation rate also determines the sludge retention time in the clarifier. It must therefore be limited, primarily to avoid bulking, anaerobic conditions and uncontrolled denitrification, which negatively impact effluent quality through sludge rising to the surface. In cases where the system design leads to high concentrations of $N\text{-}N0_3^-$ above 8 $mg.l^{-1}$, denitrification is highly likely at the bottom of the clarifier. This can then be limited by higher recirculation rates.

The recirculation rates commonly used are a compromise between the different functions. They are around 150% for the peak dry weather flow (Qptd) and 100% for the peak wet weather flow (Qptw).

The surface area of the treatment facilities is determined by the corrected volume, which in turn depends on the sludge concentration and the sludge index. These two parameters vary depending on the type of treatment. The clarifier's dimensions will also depend on these treatments.

3.6.4.4. *Concentration of biomass in the aeration tank*

The suspended solids (SS) concentration of the sludge entering the clarifier depends on optimizing the volumes of the aeration tank and the clarifier itself. Increasing the concentration in the aeration tank leads to a reduction in its volume, but this concentration is significantly reduced by increasing the required surface area of the clarifier and therefore its volume. This optimization does not yield the same concentration results for high- and low-load processes. The impact of concentration has a greater effect on the total volume of both stages (aeration and clarification) for extended aeration.

In the case of low loads, increasing the concentration in the aeration tank is more beneficial. Concentrations there are higher.

In the case of low loads (less than 0.15 kg of BOD.kg^{-1} VSS.d^{-1}) and when phosphorus treatment is required (combined or physicochemical treatment), higher concentrations are most often observed due to the increased sludge production, which requires a greater mass of sludge in the aeration tank to maintain a constant sludge age.

Table 3.6 shows the maximum recommended concentrations in the aeration tank according to the treatment type.

	Extended aeration	Low loading	Medium loading	High loading
Maximum suspended solid concentration in the aeration tank (g.L^{-1})	4.5-5	4	3	2

Table 3.6. *Maximum recommended concentrations in the aeration tank depending on the treatment type*

3.6.4.4.1. Sludge index

Depending on the type of treatment of raw wastewater (without primary settling), the reference sludge indices indicated can be used for determining the size of the clarifier, in the absence of any aggravating factors, as shown in Table 3.7.

Sludge concentration in the aeration tank (g.L^{-1})	Sludge index (mL.g^{-1})	Corrected volume (mL)	Maximum upward velocity (m.h^{-1})
4.5	150	675	0.7
3	150	450	1.07
5	–	900	0.45

Table 3.7. *Sludge concentration and upward velocity*

In the presence of physicochemical treatment with co-precipitation in the aeration tank, dephosphation allows a gain of 40 points on the sludge index.

	($ml.g^{-1}$)	TSS ($g.L^{-1}$)	Corrected volume ($ml.L^{-1}$)
WWTP without primary treatment:	130	4.5	600
– with biological dephosphatation;	130	4.5	600
– with additional precipitation;	120	5	600
– with simultaneous precipitation.	120	5	600
WWTP with primary treatment:	170	3.5	600
– with biological dephosphatation;	170	3.5	600
– with additional precipitation;	150	4	600
– with simultaneous precipitation.	150	4	600

Table 3.8. *Sludge indices according to the treatment type*

In the case of biological reactors after primary settling, the reference sludge indices increase by 50 points due to the absence of the particulate fraction that weighs down the sludge. For a rough estimate, Table 3.8 provides some additional references.

EXAMPLE.–

An activated sludge wastewater treatment plant is sized to treat carbon and nitrogen (total nitrogen discharge < 15 $mg.L^{-1}$). The incoming water is highly diluted. The size of the fully mixed aeration tank is based on a sludge concentration equal to 4.5 $g.L^{-1}$.

When determining the size of the clarifier, the reference index used is 150 $ml.g^{-1}$ of TSS. We adopt an upward velocity of 0.7 $m.h^{-1}$ for the initial sludge concentration; the effective surface area of the structure (excluding the Clifford and water recovery chutes) is obtained by dividing the peak flow rate by this upward velocity.

Given the sludge index (SI = 150 $ml.g^{-1}$), the concentration of recirculated sludge is limited to 8 $g.L^{-1}$ (1,200/SI), which gives a limiting thickening factor of 2. A thickening factor of 2 requires a recirculation rate of 200%. Lower rates result in sludge accumulation in the structure during peak demand.

3.6.4.4.2. Material balance

$$(Qe + Qr)\, Xa = Qr.Xr$$

We thus have:

$$(Qe + Qr).Xa - Qr.Xr = \text{flow of TSS stored}$$

	Unitary network
Peak duration	24 h
Stored TSS flow (kg.h^{-1})	(0.7 x 2 x 4.5) – (0.7 x 8) = 0.7 kg TSS.m^{-2}.h^{-1}
Storage height to cope with the peak	0.7 kg TSS.m^{-2}.h^{-1} x 24 h/8 g.L^{-1} = 2.10 m

Table 3.9. *Calculating the sludge storage height*

The concentration of settled sludge can be obtained using the equation:

$$Xb = \frac{1000}{SI}\sqrt[3]{ts}$$

where ts is the retention time in the clarifier (h).

3.6.4.5. *Volumetric loading at the tank bottom*

The volumetric loading at the tank bottom is given by the relation:

Volumetric loading at the tank bottom (l.m^2.h^{-1}) = SI (ml.g^{-1}) x concentration (g.L^{-1}) x upward velocity (m.h^{-1})

The higher the volumetric loading at the tank bottom, the greater the quantity of water that must be removed from the tank bottom through the sludge to the upper layers, which limits sludge settling (hindered settling).

3.6.4.6. *Mass loading at the tank bottom*

This is the total suspended solids to be settled per unit of time and per unit of surface area (kg TSS.m^{-2}.h^{-1}). It is obtained by taking into account the peak flow rate plus the flow rate recirculated by the sludge concentration at the structure inlet and brought back to the surface.

Activated sludge	**Mass flux loading at the tank bottom (kg TSS.m^{-2}.h^{-1})**
Medium loading	8–8.5
Prolonged aeration	6–6.5

Table 3.10. *Mass flux loading at the base depending on the type of treatment*

The following equation is proposed to calculate the clarifier area (S) from the bottom loading ($kg.m^{-2}.h^{-1}$):

$$S(m^2) = \frac{(1 + R)QeXa}{\text{mass loading}}$$

where R is the recirculation rate (Qr/Qe), Xa is the TSS concentration ($kg.m^{-3}$) and Qe is the inlet flow rate ($m^3.h^{-1}$).

	Very low loading		Low loading		Medium loading	High loading	Very high loading
Activated sludge supply	Raw wastewater	Primary clarified effluent	Raw wastewater	Primary clarified effluent	Primary clarified effluent	Decanted and diluted wastewater	Raw wastewater
TSS activated sludge conc. ($g.L^{-1}$)	5–6	3.5–5	4–5	3–4.5	3–4.5	2–3.5	2–3
TSS target output clarifier ($mg.L^{-1}$)	10–20	15–20	15–25	20–30	40–60	50–70	10–20
Sludge index ($ml.g^{-1}$)	90–120	120–170	110–150	135–200	135–200	170–300	50–90
Corrected volume ($ml.L^{-1}$)	450–720	420–850	440–750	405–900	405–900	340–1,050	100–270

Table 3.11. *Sludge indices and corrected volume for different loadings*

3.6.4.7. *Sludge retention time*

The frequency with which the scraper bridge moves must limit the sludge retention time and prevent turbulence in the sludge blanket. Peripheral velocities of approximately 5 $cm.s^{-1}$ are often used for structures approximately 20 m in diameter. For larger structures, the time for a complete rotation must be less than the sludge retention time. The retention time is a function of:

– the fermentable nature of the sludge:

- less than 60 minutes in prolonged aeration,

- less than 45 minutes for an average load;

– the risk of denitrification depending on the sludge quality:

- less than 20 minutes for an average load at high temperature,

- less than 45 minutes for prolonged aeration.

Settling tanks must retain sufficient activated sludge from aeration tanks to meet effluent quality standards. The activated sludge must be thickened so that the settled sludge can be continuously returned to the aeration tanks without accumulating in the settling tanks. There must be sufficient space for storing the sludge removed from the aeration tanks in case of rain.

The sludge settles and concentrates at the bottom of the clarifier, where it is scraped off and thickens in the sludge hopper. The clarified effluent is collected through chutes and discharged via a pipe. From the sludge hopper, the settled sludge is pumped back to the aeration tank as recircled sludge.

3.7. Inclined pipe clarifiers

There are two types of conventional lamellar clarifiers, distinguished by the direction of water flow. When the water flows (downward) in the same direction as the flocs, the clarifiers are co-current. When the water flows in the opposite direction to the flocs, the clarifiers are counter-current. Therefore, co-current systems facilitate floc removal, but short circuits are often observed. Counter-current systems are those in which flocculated water enters below the lamellar block. This results in the sludge crossing paths and sliding along the plates towards the bottom of the clarifier. The water is then discharged from the top and collected through chutes. Clarified water is obtained in the area where the floc concentration is lowest, resulting in clearer water. Furthermore, counter-current systems require a simpler design than co-current systems. Controlling the upward hydraulic flow is easier than controlling the downward flow in co-current systems. This compensates for the difference in projected surface area between the two systems.

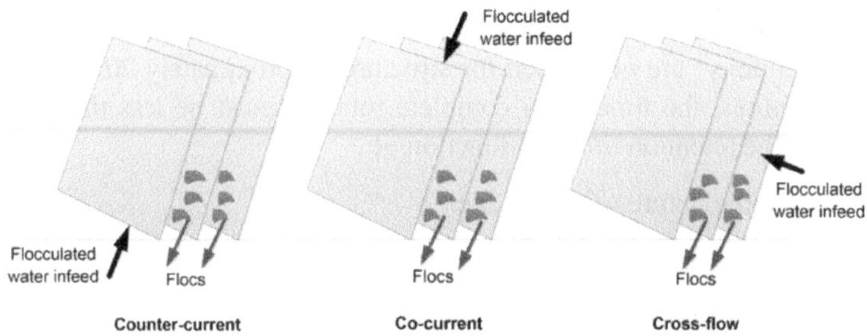

Figure 3.26. *Different flow systems for lamellar blocks*

There are also cross-flow lamellar clarifiers in which the floc flow is perpendicular to the direction of the water. The counter-current design is often preferred due to its simpler construction.

Veolia lamellar clarifiers are divided into several categories:

– with or without a sludge blanket;

– with or without weighted flocking.

3.7.1. *Theory and principle*

While the same theory used for horizontal clarifiers is applied to lamellar clarifiers, the parameter related to the presence of lamellae must be taken into account.

A lamellar module is a system of plates with inclined surfaces, whether flat or corrugated, stacked close together.

A lamellar clarifier is considered to be a compilation of many elemental clarifiers, each of which is defined by the volume between two settling plates.

Thus, two back-to-back plates form a module composed of hexagonal sections. Flocculated water is distributed beneath the modules through an orifice that extends across the entire width of the device. The water flows upwards between the plates in the opposite direction (counter-current) to the flocs, which settle on the plates and slide downwards under the influence of gravity.

The space between each lamellar plate represents a modular clarifier whose Hazen velocity (related to the cutoff threshold) is characterized by the ratio between the incoming flux and the projected surface area of the lamellar plate.

It is expressed using:

$$v\ (m.h^{-1}) = Q/Sl \cos \alpha$$

where Q is the flow rate ($m^3.h^{-1}$), Sl is the surface area of the lamellar plates (m^2) and α is the angle of inclination of the lamellar plates relative to the horizontal.

The depth of the clarifier has no influence on the cutting capacity provided that the deposited material is removed as it accumulates and does not disrupt the incoming water flow.

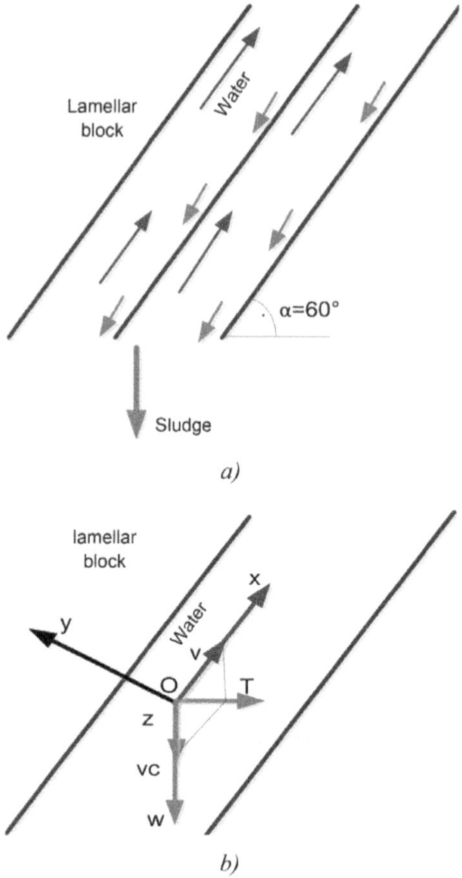

Figure 3.27. *a) Principle of counter-current operation; b) spatial representation of the liquid–solid separation*

The combined use of a high surface flow rate coupled with inclined plates (60°) allows for continuous self-scraping. There is very little floc accumulation on the plates. Let us define the coordinate system Ox, Oy, Oz, Ow, where:

– Ox is parallel to the lamellar plates.

– Oy is perpendicular to the lamellar plates.

– Oz is perpendicular to plate xOy.

– Ow is perpendicular to xOy and is related to the direction of floc deposition on the plate.

The water moves along the x-axis (clarified water outlet) with a circulation velocity between two plates (see Figure 3.27) and the flocs are subjected to a Hazen velocity (vertically along Ow) corresponding to the orientation of floc deposition and settling. The angle α is therefore determined so that the deposited floc can slide to the bottom without accumulating in any particular spot on the plates. Hence, the total projected surface area is calculated so that the laminar flow (Reynolds number < 500 and Froude number > 10^{-5}) allows gravitational forces to slide the flocs down the clarifier.

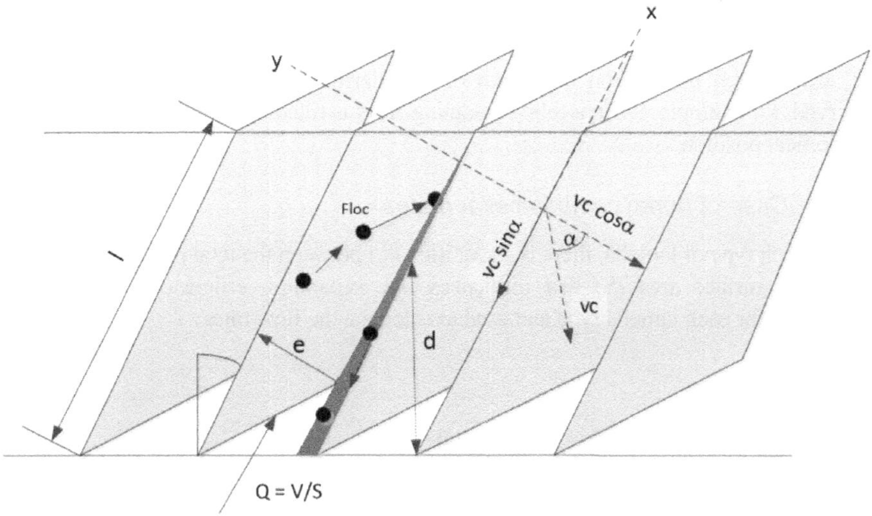

Figure 3.28. *Floc routes on lamellar plates*

Figure 3.28 shows the floc behavior between two plates. α is the angle of inclination of the plates (or lamellae), v is the surface velocity ($m.h^{-1}$), S is the clarifier surface area (m^2), Q is the cross-flow rate ($m^3.h^{-1}$), e is the gap between the plates (m), d is the vertical distance between two plates (m), vs is the floc settling rate ($m.h^{-1}$), H is the straight length of a plate (m) and l is the projected length of a plate (m).

When a particle moves along the lamellae, the time required for the particle to move through the lamellae is L/(v − vs sinα). The time required to settle on the lower lamella is H/vs cosα.

Thus, the smallest particle that can be removed by the lamellar clarifier is determined by the settling rate:

$$vs = v\frac{H}{L\cos\alpha + H\sin\alpha}$$

– If $\alpha = 0$, we find that vs = vH/L for horizontal clarifiers.

– If $\alpha = 90°$, vs = v. In this case, the floc settling velocity must be greater than the water flow velocity through the lamellae for settling to occur. Only when vs < v are the particles removed.

3.7.2. *Sizing principles for lamellar clarifiers*

The removal of particles (flocs, suspended matter) in lamellar clarifiers depends on the surface area but not on the depth of the clarifier (Hazen, Camp). This can be observed, for example, when a tube containing flocs is tilted compared to the same tube in a vertical position.

3.7.2.1. *Case of honeycomb lamellar plates*

For each type of lamella, there is a specific ratio between the total projected area (Sp) and water surface area (S) that influences floc separation efficiency. This ratio is determined for each lamella type and used to calculate the flow rates.

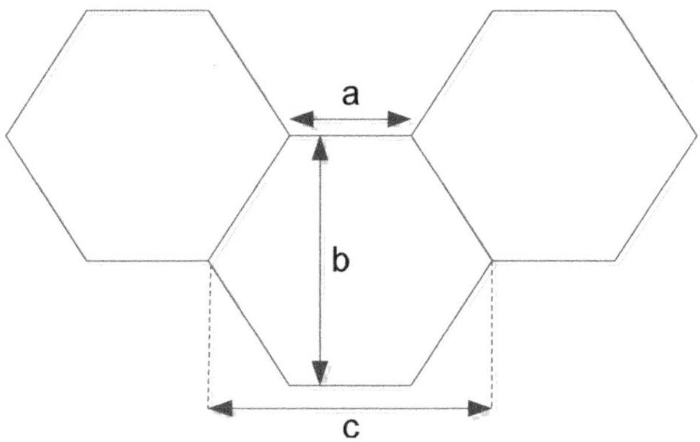

Figure 3.29. *Honeycomb lamellar structure*

We have:

– Lp: length of a lamellar plate (mm);

– a: settling width in a cell (mm);

– b: maximum orthogonal depth (mm);

– c: maximum settling width (mm);

– d: average cell width (mm) = (c + a)/2;

– b.d.: circulation surface area (mm^2);

– circulation surface area/sinα: water surface area of a cell (Sm; mm^2);

– Lx c: settling area (mm^2);

– settling area x cosα: total projected area (TPA, mm^2).

From this, we deduce the Sp/S. It is equivalent to the ratio between the surface velocity (v, $m.h^{-1}$) and the Hazen velocity (vH, $m.h^{-1}$):

Sp/S = v/vH

3.7.2.2. *Case of flat lamellar plates*

The removal of particles (flocs, suspended solids) in lamellar clarifiers equipped with flat plates is presented in Table 3.12.

Parameter	**Equation**
Hazen velocity ($m.h^{-1}$)	vH = Q/Sp
Projected surface area of a plate (m^2)	Spu = Lp.Lp.cosα
Total projected area (m^2)	Sp = n.Lp.lp.cosα
Surface velocity (S)	v = Q/S
Angle of inclination	α = 45–60°
Projected length of a plate	L = H/sinα

Table 3.12. *Dimensioning parameters for flat lamellar plates. Lp is the plate length (m), lp is the plate width (m) n is the number of plates*

Thus, determining the size of lamellar clarifiers must take into account two parameters:

– the surface velocity ($m.h^{-1}$), given by:

$$v\left(\frac{m}{h}\right) = \frac{Q}{Sm}$$

– the Hazen velocity ($m.h^{-1}$), given by:

$$vH\left(\frac{m}{h}\right) = \frac{Q}{Sp} = \frac{Q}{n.Spu.\cos\alpha}$$

where S is the water surface area (m^2), Spu is the unit plate surface area (length x width in m^2), Sp is the total projected area (m^2), e is the gap between two plates (m), L is the length of the clarifier (m), n is the number of plates and Q is the inflow rate ($m^3.h^{-1}$).

It is possible to calculate vH using (n – 1), so as not to count the plate located at the end of the clarifier, which is ineffective. However, the overall calculation remains equivalent, particularly for high-throughput plants.

The number of plates n is equal to:

$$n = \frac{L \sin\alpha}{e}$$

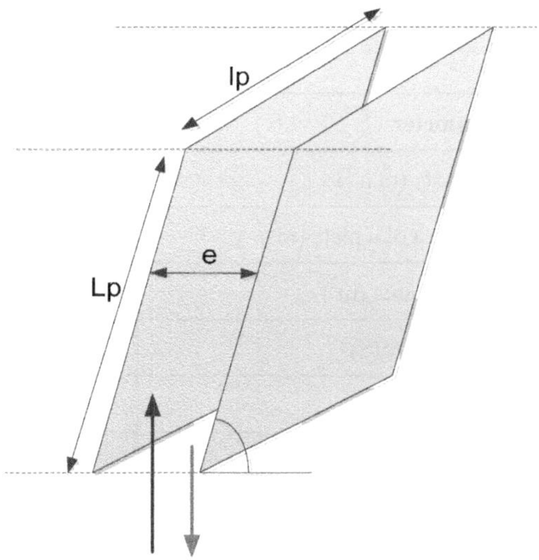

Figure 3.30. *Characteristics of flat lamellar plates*

An accurate calculation of L requires counting only the effective length containing lamellar plates, because due to their inclination, a small area at the end of the clarifier is not covered.

As with hexagonal honeycomb plates, there is a specific ratio between total projected surface area and water surface area or Sp/S, which plays a role in the efficiency of floc separation.

We have:

– Lp: length of a lamellar plate (mm);

– lp: width of a lamellar plate (mm);

– e: gap between two plates (mm);

– e.lp: circulation surface area (mm^2);

– e.lp/sinα: water surface area between two plates (Sm, mm^2);

– Lp.lp: settling surface area (mm^2);

– settling surface area x cosα: total projected surface area (mm^2).

Lamellar clarifier	**Values**	
	L = 2 m	**L = 2.5 m**
Angle of inclination	55	55
Number of waves per plate	4	4
Number of plates per module	5	5
Number of modules per pack	6	6
Number of plates per pack	30	30
Width of a lamellar plate (m)	1.030	1.030
Gap between two plates (m)	0.10	0.10
Length of a lamellar plate (m)	2.000	2.500
Orthogonal length of a module (m)	0.500	0.500
Horizontal surface length of a module (m)	0.620	0.620
Unit orthogonal surface area (m^2)	0.103	0.103
Wetted perimeter (m)	2.464	2.464
Hydraulic diameter (m)	0.167	0.167
Horizontal surface area of a module (S,m^2)	0.639	0.639
Total projected area (Sp,m^2)	5.908	7.385
Total projected area/horizontal surface area ratio	9.25	11.56
Surface velocity/Hazen velocity ratio (v/vH)	9.25	11.56
Circulation area per module (m^2)	0.515	0.515

Table 3.13. *Example of sizing based on the characteristics of the lamellae*

From this, we deduce the Sp/S. It is equivalent to the ratio between the surface velocity (v, $m.h^{-1}$) and the Hazen velocity (vH, $m.h^{-1}$):

$$Sp/S = v/vH$$

Table 3.13 gives an example of the sizing parameters.

Determining the size of the lamellar clarifier is established as follows. We have:

– flow rate = 1,000 $m^3.h^{-1}$;

– length (or height) of a plate = 2.5 m;

– plate inclination = 55°;

– orthogonal center-to-center spacing = 0.10 m;

– Hazen velocity (vH) = 0.4 $m.h^{-1}$.

The procedure is as follows:

– total projected surface area required (Sp)= 1,000/0.4 = 2,500 m^2;

– total projected surface area of a module (Spu) = 6.438 m^2;

– number of modules needed = 2,500/7.385 = 338.5 ≅ 340 modules;

– number of rows = 4;

– number of modules per row = 340/4 = 85;

– number of modules per pack = 6;

– horizontal surface area of a module = 0.639 m^2;

– total horizontal surface area (S) = 0.639*340 = 217.26 m^2;

– surface velocity (v)= 1,000/217.2 = 4.60 $m.h^{-1}$.

Ratio: Sp/S = 2,500 m^2/217.26 m^2 = 11.50

v/vH = 4.60/0.4 =11.50

If the ratio Sp/S is already known, for example, Sp/S = 11.50, we can deduce the velocity:

$$v = 11.50 \times 0.4 \ m.h^{-1} = 4.60 \ m.h^{-1}$$

$$S = 1{,}000 \ m^3.h^{-1}/4.60 \ m.h^{-1} = 217.4 \ m^2$$

$$Sp = 217.4 \ m^2 \times 11.56 = 2\ 513{,}1 \ m^2$$

With a plate length of 2.5 m, Re < 500 and the hydraulic system is in laminar flow. It is therefore possible to increase the Hazen velocity and the surface velocity, provided that the values of Re and Fr are respected.

The gap between two flat plates is generally between 3 and 5 cm with an angle of inclination of 60°. The length of the plates varies between 2 and 2.5 m. The total projected area is the area of one plate multiplied by the number of plates and by cosα. The spacing is approximately 5 cm.

3.7.2.3. *Applications for activated sludge*

The mass balance of sludge in a lamellar clarifier is written as:

$$(Qe + Qr)\ Xa = Qr\ Xr + Qe\ Xf$$

where Qe is the inlet flow rate of the activated sludge, Qr is the recirculation flow rate, Xa is the activated sludge concentration, Xr is the recycled sludge concentration and Xf is the sludge concentration at the outlet.

Comparing the inlet and recirculation values, we can estimate that Xf (30 $mg.L^{-1}$) can be neglected.

Therefore:

$$(Qe + Qr)\ Xa = Qr\ Xr$$

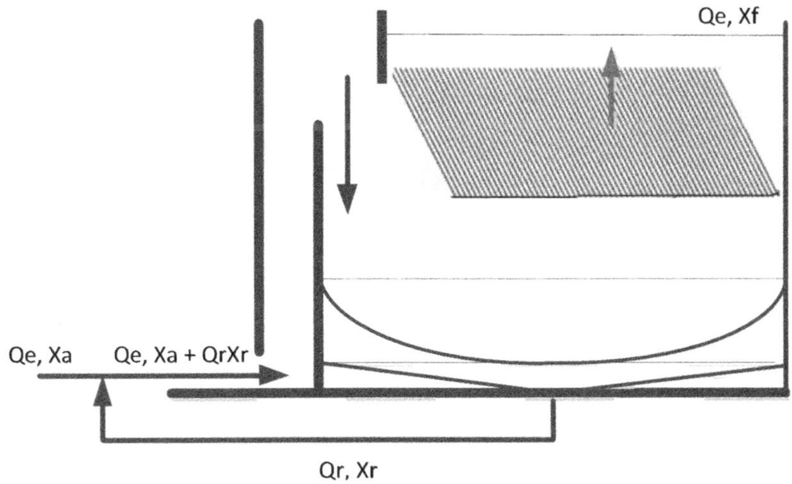

Figure 3.31. *Operating diagram of a lamellar clarifier*

The recirculation rate is the ratio between the recirculation flow rate and the inlet flow rate of the treatment plant:

$$R = 100\ \frac{Qr}{Qe}$$

The thickening factor (f) is the ratio between the recirculated sludge concentration (Xr) and the incoming sludge concentration (Xa):

$$f = \frac{Xr}{Xa} = \frac{Qe + Qr}{Qr} = 1 + \frac{Qe}{Qr}$$

whereby:

$$f = 1 + \frac{1}{R}$$

This means that for an R-value of 150%, the thickening factor is 1.65. If the R-value is 200%, f is equal to 1.5.

The permissible maximum loading is given by the expression (Canler 1994):

$$Ch = 1.2512 \text{ Exp } (-0.0031 \text{ Vc})$$

Ch is expressed in $m^3.m^{-2}.h^{-1}$. Vc is the corrected volume ($ml.L^{-1}$), with:

$$Vc = Xa \ (g.L^{-1}) \ SI \ (ml.g^{-1})$$

Thus, for an SI of 150 and a TSS concentration of 3 $g.L^{-1}$, the corrected volume is 450 $ml.L^{-1}$. The limiting hydraulic head is deduced from the previous relationship. We obtain Ch = 0.31 $m.h^{-1}$.

According to the manufacturers, more optimistic formulas can be proposed, depending on the clarifier's geometry, the flow distribution and the turbulence within the structure. The following relationship is proposed:

$$Ch = 1.45 \text{ Exp } (-0.0031 \text{ Vc})$$

At activated sludge wastewater treatment plants, the corrected volumes are generally between 450 and 600 $ml.L^{-1}$ for concentrations Xa respectively equal to 3 $g.L^{-1}$ and 4 $g.L^{-1}$ and an average SI value of 150 $ml.L^{-1}$.

This results in respective hydraulic heads of 0.36 $m.h^{-1}$ and 0.22 $m.h^{-1}$. As is generally the case, the biological tank operates with an Xa value of 3 $g.L^{-1}$. The hydraulic charge proposed by the manufacturer is 0.36–0.40 $m.h^{-1}$. Higher values are also given for lower Vc values.

In the case of a lamellar clarifier, the retained surface area is expressed as the total projected surface area (Sp) and corresponds to the sum of the projected surface areas of the plates. The following equation is proposed:

$$\text{Maximum upward rate } (m.h^{-1}) = 1.2512 \text{ Exp } (-0.0193 \text{ corrected volume})$$

The manufacturer Veolia offers Multiflo® lamellar settling tanks with a number of pieces of equipment to ensure the optimal operation of its clarifiers:

– an air washing system for cleaning the slats;

– regulating valves at the inlet of each clarifier to ensure even distribution;

– telescopic manual valves to adjust the height of the sludge blanket.

3.8. Clarifier malfunctions

The quality of effluent from a clarifier is affected by the geometry of the clarifier, the hydraulics, the characteristics of the solid matter (flocs) to be settled and the sludge concentration, as well as the sludge flow rates and extraction rates.

Evaluating the performance of a clarifier includes two important parameters: its treatment capacity and the effluent quality. The tank capacity is measured and expressed in terms of hydraulic head and sludge load at the bottom of the tank. A balance is established based on the energy balance:

$$Qe\ Xa = Qr\ Xb + Qs\ Xs$$

where:

– Qe: input flow rate ($m^3.h^{-1}$);

– Xa: TSS concentration at the outlet of the biological tank ($g.L^{-1}$);

– Qr: flow rate of extracted sludge ($g.L^{-1}$);

– Xb: extracted sludge concentration ($g.L^{-1}$);

– Qs: clarified effluent flow rate ($m^3.h^{-1}$);

– Xs: TSS concentration in the clarified effluent ($mg.L^{-1}$).

When the clarifier operates correctly according to the imposed flow rates and loads, these parameters are considered optimal for the operation of the structure.

In some cases, under conditions of peak hydraulic load or load at the tank bottom, the clarifier's operation is not satisfactory. The problem may lie in the settling and thickening of the sludge, which would explain why the clarifier's storage capacity is exceeded. The sludge accumulation rate depends on factors such as sludge compression, loading conditions, storage, the sludge removal equipment and hydraulic short circuits.

The main performance indicators of a clarifier are the quality of the clarified effluents, the sludge compression and the hydraulic behavior of the clarifier during peak flow. Sludge compression is assessed by monitoring the level of the sludge blanket and the concentration of the extracted sludge.

The hydraulic behavior of the clarifier can be evaluated by considering the following elements:

– the relationship between the level of the sludge blanket and the incoming solid concentration;

– the impact of the hydraulic load of the clarifier (and the solids load) on the quality of the outlet effluent and on the sludge blanket;

– variations in flow regimes with respect to the sludge blanket;

– the distribution of sludge within the structure.

Other types of malfunctions include:

– *Filamentous bacteria*: the excessive presence of filamentous bacteria (bulking and foaming) severely limits the hydraulic capacity of the clarifier and can lead to sludge release into the natural environment.

– *Unstable floating matter*: unstable floating matter is characterized by the presence of sludge particles not associated with filamentous bacteria. The origin of the accumulation of floating matter is threefold:

- *Denitrification*: denitrification occurs mainly in treatment plants that only process carbon pollution, where nitrification occurs spontaneously. The causes are an underload of organic matter leading to overaeration, making oxygen available for nitrification, and high temperatures that promote rapid growth of nitrifying bacteria. Better control of aeration, anoxia and mixing levels helps prevent denitrification in the clarifier. Increased recirculation from the clarifier to the aeration tank can limit denitrification by reducing the retention time of the clarifier sludge.

- *Insufficient degassing (aeroflotation phenomenon)*: this malfunction particularly affects fine bubble aerators with high diffuser immersion depths (greater than 7 meters). In this case, atmospheric nitrogen supersaturation can occur in the aeration tank. The arrival of sludge into the clarifier causes nitrogen degassing, and a phenomenon similar to biological nitrification occurs. To mitigate this problem, an efficient degasser is essential.

- *Sludge fermentation*: sludge fermentation is caused by the presence of methane, a substance generated by anaerobic reactions in the degradation of organic matter. The main causes are inadequate scraping of the tank bottom and dead zones (under the internal clarifier chute, on the scraper bridge supports, on the central well or the recirculation well), insufficient agitation in certain parts of the tank and significant underaeration. These types of unstable floating matter are easily broken down by rain.

– *Control of sludge concentration*: poor management of sludge concentration can lead to an abnormal increase in the sludge stockpile in the aeration tank, and subsequently in the clarifier. The consequences are excessive electricity consumption (due to sludge respiration) and the risk of sludge escaping from the outlet. This malfunction can occur when the conditions for removing excess sludge do not allow for sufficient extraction of the sludge produced. This situation is particularly noticeable during winter or when sludge spreading is not possible.

3.9. Technologies

3.9.1. *Clariflo® clarifier (Veolia)*

Settings	Circular shape
Water level (m)	4–5.5
Sludge concentration (g.L^{-1})	4–5.5
SI (ml.L^{-1})	< 150
Average upward rate (m.h^{-1})	1–1.1
Maximum upward rate (m.h^{-1})	2
Average mass loading (kg.m^{-2}.h^{-1})	8
Maximum mass loading (kg.m^{-2}.h^{-1})	10–12

Table 3.14. *Operating and design characteristics*

The peripheral sludge feed is split in two. The feed chute has a constant width (0.5 m) and variable height. The clarified water is returned via a notched weir with triangular teeth. The rotating bridge that recirculates the sludge and floating matter has a speed of 3.5 m.min^{-1} and rotates clockwise. The orifices are equally spaced by 1.55 m. Their diameter increases from upstream to downstream. The maximum upward velocity is 2 m.h^{-1}, and the average speed is between 1 and 1.1 m.h^{-1}. The water level is 4–5 m (see Figure 3.32A).

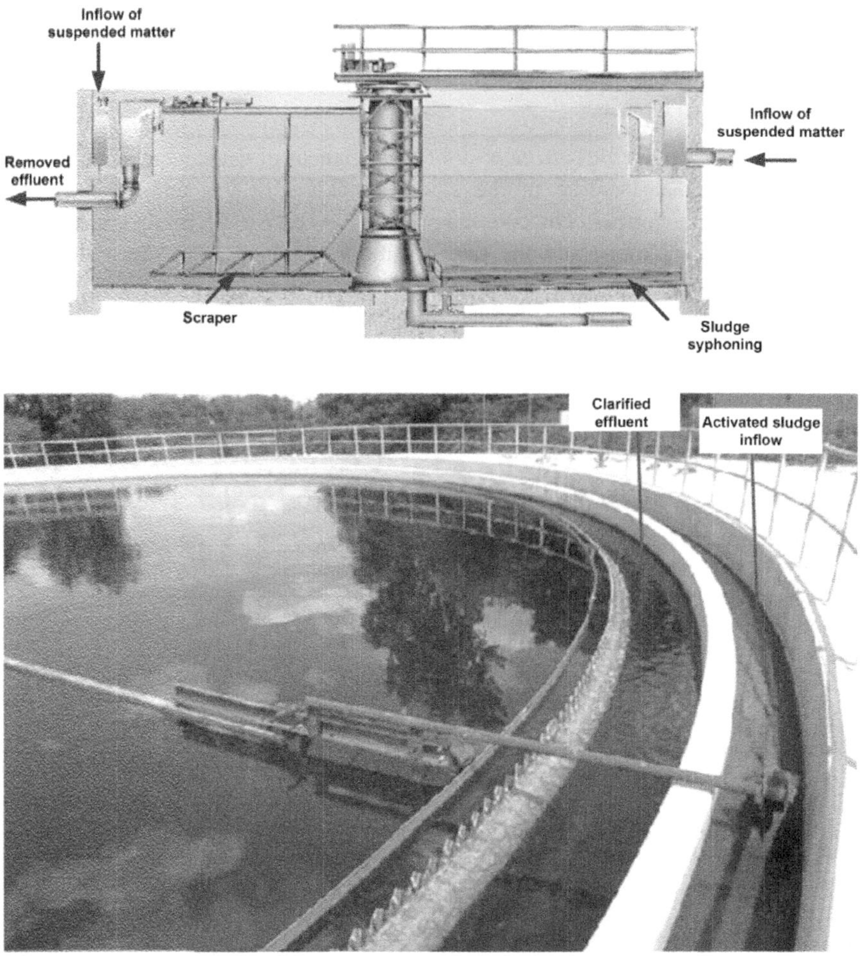

Figure 3.32A. *Clariflo® clarifier profile (Veolia). For a color version of the figure, see www.iste.co.uk/gaid/wastewaterengineering3.zip*

Table 3.14 summarizes the Clariflo® operating and design characteristics.

The main advantages of this clarifier are:

– higher surface loading;

– fewer clarifiers;

– larger permitted diameter;

– lower retention time for sludge;

– ideal for the biological removal of phosphorus;

– limited risk of uncontrolled denitrification;

– limited risk of sludge fermentation at high water temperatures.

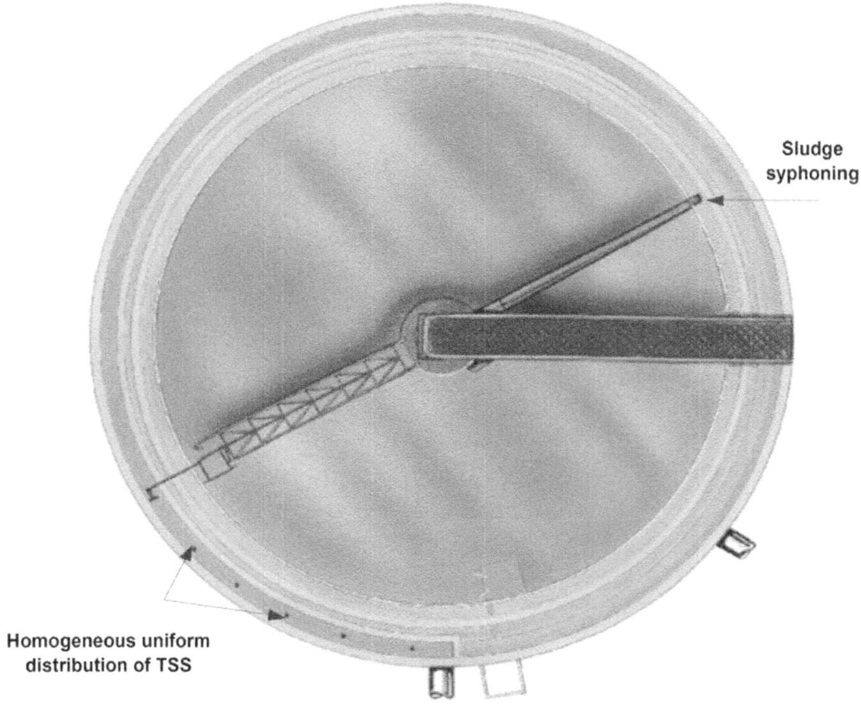

Figure 3.32B. *Top view of the Clariflo® clarifier (Veolia). For a color version of the figure, see www.iste.co.uk/gaid/wastewaterengineering3.zip*

3.9.2. *Rim-Flo® clarifier*

The Rim-Flo® clarifier is a circular, peripherally fed structure that uses a dynamic combination of flow and settling principles to achieve very good removal of suspended solids. Unlike conventional circular clarifiers, Rim-Flo® features a peripheral inlet for the biological mass from a biological tank. This peripheral feed design promotes a tangential flow pattern, creating a smooth and uniform spiral within the clarifier tank.

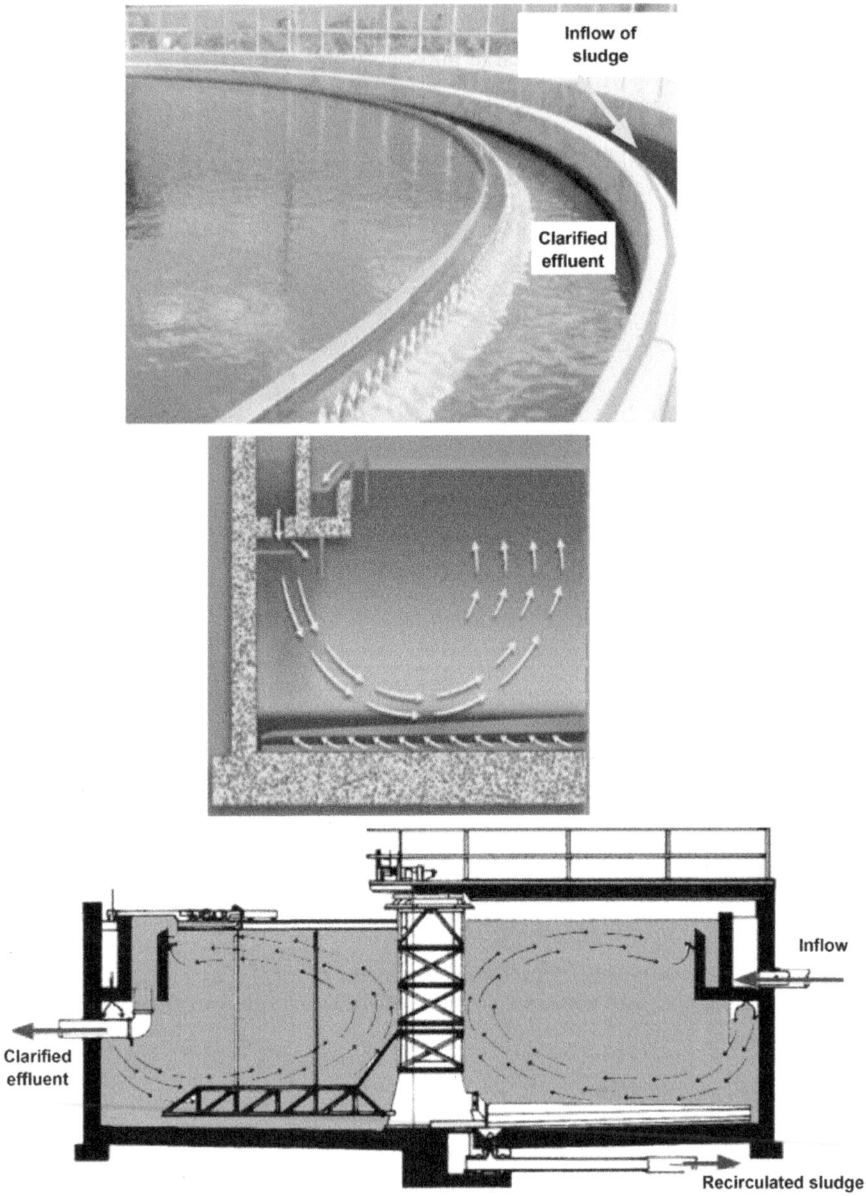

Figure 3.33. *Rim-Flo® clarifier*

Peripheral supply plays an important role in the efficiency of the Rim-Flo®. By entering the clarifier tangentially, the sludge mass creates a uniform spiral flow that minimizes short circuits and turbulence. This flow allows for maximum settling time and efficient removal of suspended solids.

The clarifier's circular design further enhances process efficiency by providing a continuous flow path for the settled wastewater, thus minimizing the risk of dead zones where sludge can accumulate.

The spiral motion promotes the gradual settling of solids at the center of the structure, facilitating sludge collection.

This technology offers several advantages compared to traditional clarifiers:

– sludge removal: the peripheral supply and circular design optimize settling time, leading to significant removal of suspended solids;

– reduced short circuit: uniform flow rate minimizes short circuits;

– reduced energy consumption: this flow dynamic reduces the need for mechanical mixing and results in reduced energy consumption.

3.9.3. *TowBro® scraper*

This is a component of a sludge removal system using suction. This system ensures the effective and efficient removal of sludge from circular clarifiers.

A TowBro® system is a system equipped with a rotating hydraulic suction header and a drag arm. It sweeps the bottom of the circular clarifier, collecting sludge through the suction head. The drag arm moves along the perimeter of the clarifier.

The TowBro® system offers several advantages over traditional sludge disposal methods:

– Efficient sludge removal: the continuous suction action guarantees excellent sludge removal.

– This system is automated, minimizing the need for manual labor.

– The gentle suction action minimizes disturbance to the settled sludge, resulting in higher quality sludge for further treatment.

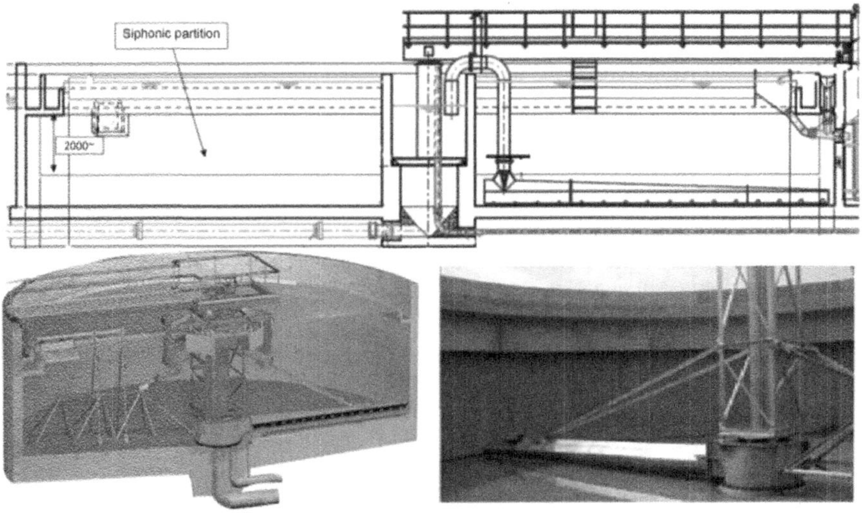

Figure 3.34. *TowBro® clarifier equipment. For a color version of the figure, see www.iste.co.uk/gaid/wastewaterengineering3.zip*

3.9.4. *Lamellar settling tank: Multiflo® (Veolia)*

Multiflo® is a lamellar clarifier designed to separate the liquid and solids of activated sludge. As in horizontal clarifiers, the water flows upwards and exits at the top, while the flocs settle by gravity. The difference between these two types of settling tanks is that in lamellar settling tanks, the flow is directed towards inclined plates, which promotes homogeneous hydraulic distribution and avoids unstable flows that disrupt settling.

Furthermore, the inclination of the plates (60°), their lengths (2–2.5 m), an optimal spacing between them and the projected surface area of each can reduce the floc travel distance and make them reach the plates quickly. They then clump together on the surface and slide down the clarifier.

Unlike traditional lamellar clarifiers, which are solely dedicated to phase separation, thickening clarifiers such as the Multiflo® allow, within the same structure, the separation of the two phases and the safe concentration of the sludge. This technology incorporates numerous features that ensure good performance:

– construction of an advanced settling zone of sufficient size to allow the separation of heavy particles;

– the area under the plates is increased to increase the sludge retention time, capture a significant portion of the suspended solids not retained in the first zone and prevent sludge resuspension by the incoming current;

– at the same time, the height of the plates is limited, eliminating any risk of sludge stagnation between the plates;

– the size of the settling tank equipped with lamellae (horizontal surface) depends on the type of floc and ensures an upward velocity lower than that of sludge flowing in the outgoing liquid;

– the spacing and length of the plates are a compromise between the required quality of the clarified water, the mass of sludge likely to be deposited on the plates and the adhesion of this sludge;

– the sludge compartment is broken up, allowing for the removal of water pockets and the remixing of fresh sludge with that already in the settling tank, thus preventing the degradation of the deposits;

– the sludge collection takes place at a single point;

– controlled recirculation of the deposited sludge allows for an increase in the initial concentration and therefore the final concentration for the same retention time;

– maintaining a retention time lower than the maximum value for the given sludge type, in order to prevent its degradation;

– extraction of thickened sludge by syncopation (rate-duration), with extraction times depending on the mass of sludge produced, thus avoiding the extraction of excessively liquid sludge during the night;

– maintaining a safety zone between the sludge blanket and the lamellae to allow optimal use of the area reserved for sludge and thus promote thickening.

The manufacturer provides a nomogram to define the mirror velocity (vm) as a function of the corrected volume (Vc) and for a desired suspended solids concentration in the outlet effluent.

Table 3.15 summarizes some target values used for sizing the Multiflo®-type lamella clarifier.

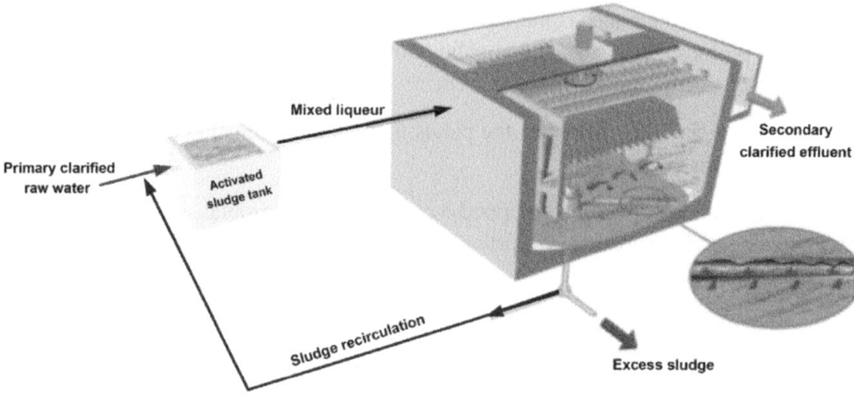

Figure 3.35. *Multiflo® lamellar clarifier*

Outlet TSS concentration (mg.L^{-1})	Corrected volume (ml.L^{-1})	Surface velocity (m.h^{-1})	Maximum sizing of vm
30	450	3.80	4.50
20	450	2.30	–
10	450	1.10	–
30	600	2.10	3
20	600	1.40	–
10	600	0.50	–

Table 3.15. *Determining the surface velocity as a function of the corrected volume for the Multiflo® lamellar clarifier*

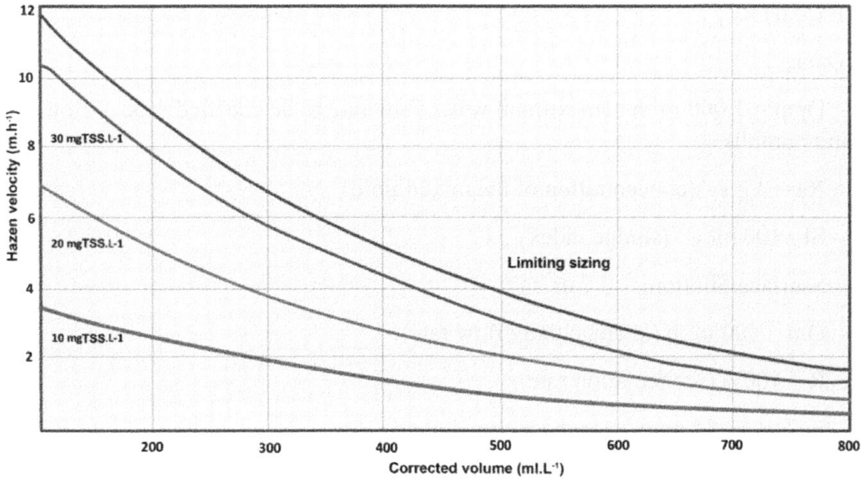

Figure 3.36. *Determination of the surface velocity as a function of the corrected volume for the Multiflo® lamellar clarifier. For a color version of the figure, see www.iste.co.uk/gaid/wastewaterengineering3.zip*

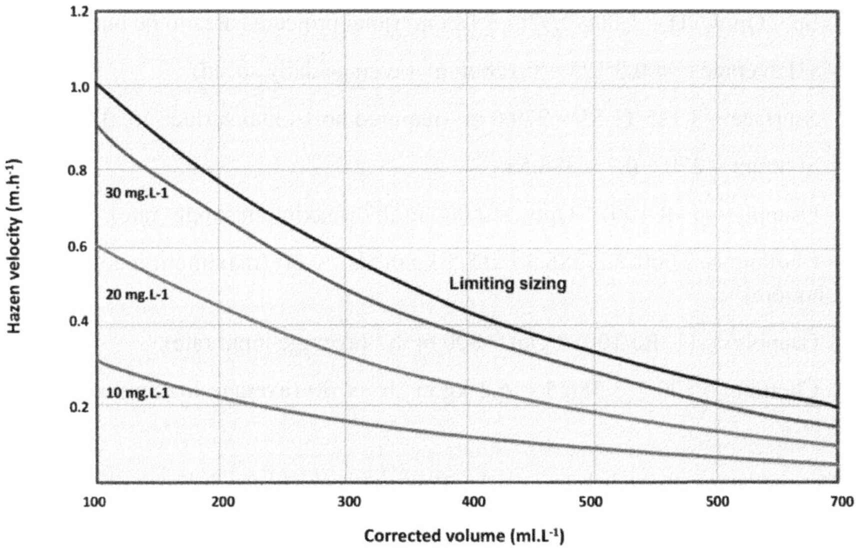

Figure 3.37. *Determination of the Hazen velocity as a function of the corrected volume for the Multiflo® lamellar clarifier. For a color version of the figure, see www.iste.co.uk/gaid/wastewaterengineering3.zip*

EXAMPLE.–

Data:

– Qptp = 1,000 $m^3.h^{-1}$ (maximum water flow rate to be clarified = peak flow rate during rainfall).

– Xa = 3 $g.L^{-1}$ (concentration of the mixed fluid).

– SI= 100 $ml.g^{-1}$ (sludge index).

– Ssurface/Sbottom = 0.7 $m^2.m^{-2}$.

– Qm = 400 m^3/h (average daily flow rate).

– R = 100% (recirculation rate).

– Sp/S = 11.55 $m^2\ m^{-2}$ for the inclined plates.

– Desired outlet TSS concentration: 30 $mg.L^{-1}$.

Calculations:

– Vc = 3 * 150 = 450 $ml.L^{-1}$ (corrected volume).

– vH (see Figure 3.37) = 0.32 $m.h^{-1}$ (safe) and 0.38 $m.h^{-1}$ (Qptw velocity limit).

– Sp = Qptw/vH = 1,000/0.32 = 3 125 m^2 (total projected area to be built).

– vH average = 400/3 125 = 0.128 $m.h^{-1}$ (average daily speed).

– Ssurface = 3,125/11.55 = 270.6 m^2 (required horizontal surface area).

– Sbottom = 270.6/0.7 = 386.5 m^2.

– Qsupply= (1+R/100) * Qptw = 2 000 $m^3.h^{-1}$ (maximum supply rate).

– Cbottom = 2,000 * 3/386.5 = 15.5 $kg.m^{-2}.h^{-1}$ < 30 (maximum loading on the tank bottom).

– $Qsupply_2$ = (1+Rc/100) * Qm = 800 $m^3.h^{-1}$ (average input rate).

– Cbottom = 800 * 3/386.5 = 6.2 $kg.m^{-2}.h^{-1}$ < 30 (average loading on the tank bottom).

Depending on the reliability of the flow rate, which can be overestimated, the limiting sizing values are, for 30 mg $TSS.L^{-1}$ at the outlet:

– vH = 0.38 $m.h^{-1}$;

– Sp = 2,631 m^2;

– S surface = 227.8 m^2;

– C bottom = 26.3 m^2 for the peak flow.

3.10. References

Adelman, M.J., Hurst, M.W., Weber-Shirk, M.L., Cabrito, T.S., Somogyi, C., Lion, L.W. (2013). Floc roll-up and its implications for the spacing of inclined settling devices. *Environ. Eng. Sci.*, 30(6), 302–310.

ATV-DVWK (2000). Dimensioning of single stage activated sludge plants. Report A 131F, volume 1. Hennef.

ATV-DVWK (2003). Dimensioning of single stage activated sludge plants. Report A 198E, volume 1. Hennef.

Bargiel, M. and Tory, E.M. (2013). Extension of the Richardson–Zaki equation to suspensions of multisized irregular particles. *Int. J. Mineral Process.*, 120, 22–15.

Blazy, P., Jdid, E.-A., Bersillon, L.-L. (2013). *Décantation – aspects théoriques*. Techniques de l'Ingénieur, Paris.

Bueno-de-las-Heras, J.L., Gutiérrez-Lavín, A., Mahamud-López, M.M. (2024). Applying a novel generalized model of sedimentation to the design of settlers and thickeners: Integration and extension of the classical procedures. *Ann. Environ. Sci. Toxicol.*, 8(1), 43–56.

Bürger, R., Careaga, J., Diehl, S. (2017). A simulation model for settling tanks. *Chem. Eng. Commun.*, 204(11), 1270–1281.

Canler, J.-P. (1994). La décantation lamellaire. Technical Document, FNDAE, Cemagref.

Canler, J.P. and Perret, J.M. (2005). La clarification : approche dimensionnelle basée sur le couple aération-clarification. Fondements et evolutions. Periodical Article, Office internationale de l'eau.

Canler, J.-P. and Perret, J.-M. (2008). Les clari-floculateurs. Technical Document, FNDAE, Ministère de l'ecologie, du développement, et de l'aménagement durables, Paris.

Carlsson, B. (1998). An introduction to sedimentation theory in wastewater treatment. Aquatic and Environmental Engineering Program, Uppsala University.

Cho, S.H. (1993). Settling velocity model of activated sludge. *Water Res.*, 27(7), 1237–1242.

Cho, K.H., Jung, S.H., Kim, T.G., Park, K.H. (1996). Defuzzification method for multi-shaped output fuzzy sets. *Electronics Letters*, 30(9), 740–742.

Coe, H.A. and Clevenger, G.H. (1916). Methods for determining the capacities of slime-settling tanks. *Trans. AIME*, 55, 356.

Cole, R.F. (1968). Experimental evaluation of the Kynch theory. Dissertation Thesis, University of North Carolina at Chapel Hill.

Concha, F. and Bustos, M.C. (1991). Settling velocities on particulate systems of Kynch sedimentation processes: Batch settling. *Int. J. Mineral Process.*, 32(3), 193–212.

Crittenden, J.C., Rhodes Trussell, R., Hand, D.W., Howe, K.J., Tchobanoglous, G. (2012). *Water Treatment Principles and Design*. John Wiley & Sons, Hoboken, NJ.

Davis, M.L. (2010). *Water and Wastewater Engineering: Design Principles and Practice*. McGraw Hill, New York.

Duchène, P. (1988). Qualité des boues activées et dimensionnement des décanteurs secondaires. Technical Document 14, March.

Ekama, G.A. and Marais, P. (2002). Hydrodynamic modelling of secondary settling tanks. Water Resources Group Report 835/1/02, Part 1, Department of Civil Engineering, University of Cape Town.

Ekama, G.A., Barnard, J., Gunthert, F., Krebs, P., McCorquodale, J.A., Parker, D.S., Wahlberg, E.J. (1997). *Secondary Settling Tanks: Theory, Modelling, Design and Operation*. International Association on Water Quality, Richmond.

Fair, G.M., Geyer, J.C., Okun, D.A. (1968). *Water and Wastewater Engineering. Volume 2. Water Purification and Wastewater Treatment and Disposal*. John Wiley & Sons, Hoboken, NJ.

Fitch, B. (1966). Current theory and thickener design. *Ind. Eng. Chem.*, 58, 18–28.

Fitch, B. (1979). Sedimentation of flocculent suspensions: State of the art. *AIChE*, 25(6), 913.

Fitch, B. (1983). Kynch theory and compression zones. *Aiche Journal*, 29(6), 940–947.

Fitch, B. (1995). Flow path effect on sedimentation. *Sewage & Industrial Wastes*, 28(1), 1–15.

Fitch, B. and Adrian, D.D. (1974). Gravity thickening of water treatment plant sludge. In *94th Annual Conference*, Boston.

Forsell, B. and Hedström, B. (1975). Lamella sedimentation: A compact separation technique. *Water Pollut. Control Federation*, 47(4), 834–842.

Galvin, K.P. and Waters, A.G. (1985). The influence of flocculant on the three sedimentation zones in continuous thickening. In *Proc. 3rd Aust. Coal. Prep. Conf.*, 291.

Galvin, K.P. and Waters, A.G. (1987). The effect of sedimentation feed flux on the solids flux curve. *Powd. Tech.*, 53, 113.

Hassett, N.J. (1958). Design and operation of continuous thickeners. *Ind. Chem.*, 34, 116.

Hazen, A. (1904). On sedimentation. *Trans. Am. Soc. Civil Eng.*, 53, 45–71.

Imam, E., McCorquodale, J.A., Bewtra, J.K. (1983). Numerical modeling of sedimentation tanks. *J. Hydraulic Eng.*, 109(112), 1740–1754.

Jernqvist, A.S.H. (1965). Experimental and theoretical studies of thickeners, Part I. *Svensk Papperstidning*, 68, 506–517.

Jernqvist, A.S.H. (1966). Experimental and theoretical studies of thickeners, Part 4. Experimental results. *Svensk Papperstidning*, 69, 395–398.

Keinath, T.M. (1977). Activated sludge-unified system design and operation. *J. Environ. Eng. Div.*, 103(5), 829–849.

Kynch, G.J. (1952). A theory of sedimentation. *Trans. Faraday Soc.*, 48, 166–176.

Laquidara, V.D. and Keinath, T.M. (1983). Mechanism of clarification failure. *J. Water Poll. Control Fed.*, 55(1), 54–57.

Lavín, A.G. (1984). A contribution to the study of settlers: Pilot plant for flocculation, clarification and thickening of coal mining wastewater. PhD Thesis, University of Oviedo.

Lipke, S. and Ramalingam, K. (2013). Implementation of CFD modeling in the performance assessment and optimization of secondary clarifiers: The PVSC case study. *Water Sci. Technol.*, 68(9), 1901–1913.

Lytra, M. (2019). Hydraulics of lamella sedimentation, a study of the lamella settlers in Ringsjö water treatment plant, Sweden. Dissertation Thesis, Lund University.

Matko, T., Fawcett, N., Sharp, A., Stephenson, T. (1996). A numerical model of flow in circular sedimentation tanks. *Inst. Chem. Eng.*, 74(Part B), 197–204.

Merta, H. and Ziolo, J. (1986). Calculation of thickener area and depth based on the data of batch-settling test. *Chem. Eng. Sci.*, 41, 1918–1920.

Michaels, A.S. and Bolger, J.C. (1962). Settling rates and sediment volumes of flocculated kaolin suspensions. *I&EC Fundamentals*, 1, 24–33.

Muñiz, M., Berthiaux, H., Lavín, A.G., Bueno, J.L. (1998). Pilot-plant studies of non-steadystate continuous sedimentation: Hydrodynamics and solids distribution profiles. *Powder Technol.*, 96, 267–274.

Nadeau, C. (2016). Modélisation de l'épaississement de la boue rouge par cisaillement. PhD Thesis, Université du Québec, Chicoutimi.

Obata, E. and Watanabe, Y. (1979). Analysis of semi-continuous thickener using a batch settling curve. *J. Chem. Eng. Japan*, 12(4), 307–312.

Perret, J.-M. and Canler, J.-P. (2008). Synthèse sur les connaissances acquise sur les clarifloculateurs en traitement tertiaire. *Techniques, Sciences et Méthodes*, 103(1), 55–68.

Poinsot, L. (2002). État de l'art sur la clarification secondaire dans l'approche dimensionnelle et sur la modélisation, limité aux clarificateurs cylindriques. Thesis, University of Strasbourg.

Reyes, C., Apaz, F., Niño, Y., Barraz, B., Arratia, C., Ihle, C.F. (2022). A review on steeply inclined settlers for water clarification. *Miner. Eng.*, 38, 107639.

Rivet, P. (1981). *Guide de la séparation liquide-solide.* Société Française de Filtration, Idexpo, Cachan.

Rudman, M., Simic, K., Paterson, D.A, Strode, P., Brent, A., Sutalo, I.D. (2008). Raking in gravity thickeners. *Int. J. Mineral Process.*, 86(8), 114–130.

Rudman, M., Paterson, D.A., Simic, K. (2010). Efficency of raking in gravity thickeners. *Int. J. Mineral Process.*, 95(10), 30–39.

Sarkar, S., Kamilya, D., Mal, B.C. (2007). Effect of geometric and process variables on the performance of inclined plate settlers in treating aquacultural waste. *Water Res.*, 41(5), 993–1000.

Shirato, M., Kato, H., Kobayashi, K., Sakazaki, H. (1970). Analysis of settling of thick slurries due to consolidation. *J. Chem. Eng. Japan*, 3, 98–104.

Suttill, K.R. (1991). The ubiquitous thickener evolves into several complementary designs. *Eng. Mining J.*, February, 20–26.

Talmage, W.P. and Fitch, E.B. (1955). Determining thickener unit areas. *Ind. Eng. Chem.*, 47, 38–41.

Tiller, F.M. (1981). Revision of Kynch sedimentation theory. *AIChE*, 27(5), 823–829.

Torfs, E., Marti, M.C., Locatelli, F., Balemans, S., Burger, R., Diehl, S., Laurent, J., Vanrolleghem, P.A., François, P., Nopens, I. (2017). Concentration-driven models revisited: Towards a unified framework to moldel settling tanks in WWRFs. *Water Sci. Tech.*, 75, 539–551.

Valle Medina, M. (2019). Modélisation des décanteurs secondaires : etude en mécanique des fluides numériques de la dynamique de la sédimentation des boues activées. PhD Thesis, University of Strasbourg.

Vesilind, P.A. (1968). Influence of stirring in the thickening of biological sludge. Thesis, University of North Carolina at Chapel Hill.

Water Environment Federation (1998). *Design of Municipal Wastewater Treatment Plants*, 4th edition. Water Environment Federation, Alexandria.

Water Environment Federation (2005). *Clarifier Design*, 2nd edition. Water Environment Federation, Alexandria.

Waters, A.G. (1989). Theory and application of thickener design. In *Dewatering Technology and Practice Conference*, Brisbane.

Wilhelm, J.H. and Naide, Y. (1981). Sizing and operating continuous thickeners. *Mining Eng.*, 33, 1710–1718.

Zioło, J. (1996). Influence of the system geometry on the sedimentation effectiveness of lamella settlers. *Chem. Eng. Sci.*, 51(1), 149–153

Index

A, B, C

activated sludge, 11, 13, 21, 22, 27, 28, 30, 31, 37–40, 43, 53, 58, 72, 73, 79, 101–103, 114, 117, 122, 123, 127, 130, 131, 134, 135, 152, 153, 155, 171, 178, 193, 197–200, 207, 208, 213, 226, 229, 231, 232, 241, 242, 250
ammonia, 1, 2, 4, 5, 7, 8, 19, 52, 53, 55, 61–63, 65–67, 69, 76, 80
ammonification, 5, 6, 9, 112
bacteria, 2, 4–23, 31, 40, 42–44, 46, 51–55, 57–67, 69, 71–73, 75, 76, 79, 103, 105–109, 111, 114, 125, 129, 130, 132, 151, 159, 214, 244
  aerobic, 11, 79, 107
  anaerobic, 9, 63, 79
  annamox, 52–55, 57–63, 66, 67, 69, 71–73, 75, 76
  sulfate-
    oxidizing, 14, 15
    reducing, 14, 16
biolfilters, 125
chemical precipitation, 97, 151
clarification, 115, 152, 155, 156, 160, 171, 180, 195, 197, 199, 200, 207, 212, 228
clarifiers
  cylindrical, 221
  horizontal, 216, 219–221, 233, 236
compaction zone, 180, 208, 209

D, E

denitrification, 9–15, 17, 25–27, 30–34, 36, 37, 39–47, 49–54, 63, 65, 66, 68–70, 73, 75, 78, 80, 81, 106, 109, 112, 115, 117, 122, 123, 127, 130, 131, 200, 215, 227, 231, 244, 247
  autotrophic, 14, 40, 41, 43, 44, 49–51, 80
denitritation, 64–67, 80
dephosphatation/phosphorus removal, 36, 97, 99, 103, 104, 106, 107, 109–112, 114–117, 122, 123, 132, 140, 150, 153, 156, 225, 229
  biological, 36, 123
eutrophication, 1, 2, 34, 93–95, 100, 102, 129, 155

L, M, N

limestone, 40, 43, 49–51, 156
load(ing)
  mass, 22–24, 28, 33, 35, 214
  underload of organic matter, 244
  volumetric, 23–25, 34, 49, 50, 230

maerl, 46, 50, 51, 157
nitrates, 6, 7, 9, 10, 12–14, 16, 17, 30–32, 35, 37, 40, 42–47, 50, 53, 54, 56, 62, 63, 65, 75, 80, 95, 99, 104–107, 109, 112, 117, 122, 131, 215
nitrification, 6–11, 16–20, 22–27, 29, 30, 37–39, 52, 53, 62, 63, 65, 66, 68–70, 73, 75, 80, 106, 111, 112, 115, 123, 127, 128, 131, 244
nitritation, 7, 62–67, 69, 71, 73, 75, 76, 80
nitrites, 6–9, 12, 13, 40, 43, 44, 53–55, 57, 58, 60, 61, 63, 65, 69, 76
nitrogen, 1–6, 8–11, 13, 14, 17, 20, 21, 25–27, 29–31, 34, 37, 39, 40, 44, 51–53, 57, 61–63, 65–73, 75, 76, 78, 80, 93, 95, 102, 106, 112, 117, 123, 127, 130, 131, 151, 159, 229, 244

O, P

organic, 2, 4, 6, 51
  carbon, 11–13, 37, 41, 53, 68, 75, 76, 123
  matter, 2, 12, 23, 28, 37, 64, 80, 138, 143, 150, 155, 244
oxygen requirements, 25, 26, 53, 80, 131
phosphates, 46, 94–99, 103, 105, 110, 111, 113, 114, 117, 127, 135, 139, 140, 147–153, 155, 157–160
phosphorus, 1, 13, 20, 75, 93–107, 109–112, 114–116, 122, 123, 125–128, 130, 132–136, 140, 141, 147–151, 154–157, 159, 160, 228, 247
post-denitrification, 37, 39, 70, 160
pre-denitrification, 37, 39, 131
process
  ANAMMOX®, 65, 77
  DeAmmon, 69
  Sharon®, 67, 69, 75
  Terra-N , 69

R, S, V

rate
  growth, 13, 17, 18, 20, 21, 27–29, 41, 42, 46, 55, 58, 67, 76, 107
  recirculation, 23, 33, 114–116, 122, 132, 198, 201, 212, 227, 229, 231, 241, 254
selectors, 130–132
settling
  coalescent, 180, 209
  curve, 178, 180, 183, 185, 196, 198, 199
  free, 181, 202
  hindered, 180, 182, 193, 207, 209
sizing clarifiers, 212, 214, 220, 236, 237
sludge index, 23, 200, 210, 212–214, 225–227, 229, 254
sulfates, 14, 15, 40, 43, 45, 46, 50, 136
sulfur, 13–15, 40–52
velocity/rate
  denitrification, 30, 31, 46, 48, 49
  nitrification, 24
  settling, 174, 175, 179, 181–184, 187–191, 193, 195, 196, 198, 199, 202–208, 210, 211, 223, 235
  upward, 195, 200, 201, 210, 213–215, 225–230, 245, 251

Other titles from

iSTE

in

Ecological Science

2026

GAID Kader
Wastewater Engineering and Treatment 1: Composition, Pollutant Loads, Primary Settling, Activated Sludge
Wastewater Engineering and Treatment 2: Membranes, Hybrid Systems, Bacteria Beds and Biofilters, Biological Disks

2025

COURANT Dimitri, REBER Bernard
Deliberative Democracy and Ecological Transition: The French Citizens' Convention for Climate
(Planetary Changes, Decision, Governance Set – Volume 1)

2024

AMIARD Jean-Claude, ZERBIB Jean-Claude
Radioactive Risk for Humans
(Radioactive Risk Set – Volume 7)

AKBAR Saima, KUMAR SINGH Suraj, KANGA Shruti
Disaster Management: Preparedness and Mitigation

FLEURENCE Joël
Aquatic Biotechnologies: From Genetic Engineering to Enzymatic or Fermentation Engineering

GAID Kader
Wastewater Reuse 1: Characteristics, Uses, Applications, Filtration and Disinfection of Water
Wastewater Reuse 2: Micropollutants, Membranes and Treatment Procedures

MATAGNE Patrick
Birth of Scientific Ecology: Eugenius Warming (1841–1924)

2023

GAID Kader
Drinking Water Treatment 1: Water Quality and Clarification
Drinking Water Treatment 2: Chemical and Physical Elimination of Organic Substances and Particles
Drinking Water Treatment 3: Organic and Mineral Micropolluants
Drinking Water Treatment 4: Membranes Applied to Drinking Water and Desalination
Drinking Water Treatment 5: Calco-carbonic Equilibrium and Disinfection

2022

AMIARD Jean-Claude
Marine Radioecology
(Radioactive Risk Set – Volume 6)

GODET Laurent, DUFOUR Simon, ROLLET Anne-Julia
The Baseline Concept in Biodiversity Conservation: Being Nostalgic or Not in the Anthropocene Era

GRISON Claude, CASES Lucie, LE MOIGNE Mailys, HOSSAERT-MCKEY Martine
Photovoltaism, Agriculture and Ecology: From Agrivoltaism to Ecovoltaism

LÉVÊQUE Christian
Biodiversity Erosion: Issues and Questions

2021

AMIARD Jean-Claude
Disarmament and Decommissioning in the Nuclear Domain
(Radioactive Risk Set – Volume 4)
Management of Radioactive Waste
(Radioactive Risk Set – Volume 5)

FLEURENCE Joël
Microalgae: From Future Food to Cellular Factory

JOLY Fernand, BOURRIÉ Guilhem
Mankind and Deserts 1: Deserts, Aridity, Exploration and Conquests
Mankind and Deserts 2: Water and Salts
Mankind and Deserts 3: Wind in Deserts and Civilizations

PIOCH Sylvain, SOUCHE Jean-Claude
Eco-design of Marine Infrastructures: Towards Ecologically-informed Coastal and Ocean Development

2020

BRUSLÉ Jacques, QUIGNARD Jean-Pierre
Fish Behavior 1: Eco-ethology
Fish Behavior 2: Ethophysiology

ROSSIGNOL Jean-Yves
Climatic Impact of Activities: Methodological Guide for Analysis and Action

2019

AMIARD Jean-Claude
Industrial and Medical Nuclear Accidents: Environmental, Ecological, Health and Socio-economic Consequences
(Radioactive Risk Set – Volume 2)
Nuclear Accidents: Prevention and Management of an Accidental Crisis
(Radioactive Risk Set – Volume 3)

BOULEAU Gabrielle
Politicization of Ecological Issues: From Environmental Forms to Environmental Motives

GIRAULT Yves
UNESCO Global Geoparks: Tension Between Territorial Development and Heritage Enhancement

KARA Mohamed Hichem, QUIGNARD Jean-Pierre
Fishes in Lagoons and Estuaries in the Mediterranean 2: Sedentary Fish
Fishes in Lagoons and Estuaries in the Mediterranean 3A: Migratory Fish
Fishes in Lagoons and Estuaries in the Mediterranean 3B: Migratory Fish

OUVRARD Benjamin, STENGER Anne
Incentives and Environmental Policies: From Theory to Empirical Novelties

2018

AMIARD Jean-Claude
Military Nuclear Accidents: Environmental, Ecological, Health and Socio-economic Consequences
(Radioactive Risk Set – Volume 1)

FLIPO Fabrice
The Coming Authoritarian Ecology

GUILLOUX Bleuenn
Marine Genetic Resources, R&D and the Law 1: Complex Objects of Use

KARA Mohamed Hichem, QUIGNARD Jean-Pierre
Fishes in Lagoons and Estuaries in the Mediterranean 1: Diversity, Bioecology and Exploitation

2016

BAGNÈRES Anne-Geneviève, HOSSAERT-MCKEY Martine
Chemical Ecology

2014

DE LARMINAT Philippe
Climate Change: Identification and Projections

Printed and bound by CPI Group (UK) Ltd, Croydon, CR0 4YY

25/06/2026

14907692-0002